A Visual Analogy Guide to Chemistry

Paul A. Krieger

Grand Rapids Community College

Edited by

Dr. Sandra L. Andrews
Professor of Chemistry,
Grand Rapids Community College

MORTON
PUBLISHING

925 W. Kenyon Avenue, Unit 12
Englewood, CO 80110
800-348-3777
www.morton-pub.com

Book Team

Publisher	Douglas N. Morton
Biology Editor	David Ferguson
Editorial Assistant	Rayna Bailey
Production Manager	Joanne Saliger
Production Assistant	Will Kelley
Cover Design	Bob Schram, Bookends, Inc.
Illustration	Paul Krieger

Library of Congress Control Number: 2011942015

ISBN-13: 978-089582-835-4

10 9 8 7 6 5 4 3 2 1

Printed in the United States of America

For my wife, Lily

For your love, constant support, and endless encouragement

Without your help this book would

never have been written

And for my son, Ethan

May you never lose that sparkle in your eye and

May you always follow your passions in life

Acknowledgments

Many people contributed to this book. First, I acknowledge Doug Morton, president of Morton Publishing, for agreeing to publish my book. As an author, working with Morton Publishing has always been a pleasure. David Ferguson, my editor, for his support, patience, and thoughtfulness in giving me the extra time and the freedom to complete the book the way I originally envisioned it. Sandy Andrews, my content editor for chemistry, offered her outstanding scholarship, excellent suggestions, and attention to detail. Her numerous insightful comments greatly improved the final manuscript in countless ways. Dan Matusiak created the original glossary. Joanne Saliger and her production design team—Will Kelley and Carolyn Acheson—deserve special recognition for the terrific work they have consistently done in producing all the Visual Analogy Guides. Rayna Bailey, editorial assistant, thoroughly checked the manuscript. I also offer a special thank you to reviewers Coretta Fernandes, Lansing Community College, Felix Ngassa, Grand Valley State University, and Shawn Macauley, Muskegon Community College for their valuable feedback during the writing process. Bob Schram, graphic designer at Bookends, Inc., created a unique and engaging book cover. My wife, Lily, kindly accepted all my long working hours away from the family. Without her help this book would not have been written. Finally, for my students, colleagues, friends, and everyone else who offered suggestions, support, and encouragement—thanks to all of you. I am truly grateful.

Contents

How To Use This Book

Purpose

This book was written primarily for students in an introductory chemistry course such as an introduction to general, organic, and biochemistry; however, it will be useful for teachers or anyone else with an interest in the subject matter. It is designed to be used in conjunction with a similar chemistry textbook. What makes it unique, creative, and fun is the *visual analogy learning system*, explained below. The modular format allows you to focus on one key concept at a time. Each module is a two-page layout, sometimes consisting of a text page on the left with corresponding illustrations on the facing page. Other times, illustrations are presented on both pages to compare opposing concepts such as acids and bases. Practice problems, along with the answers, are often integrated into the modules to assess student learning. The book focuses on key concepts and content in general, organic chemistry, and biochemistry. It is presented in a two-color format, inviting you to treat it like a workbook in which you can take notes, highlight key points, and mark it up as you see fit. In addition, it offers rhymes, jokes, and study tips for mastering difficult topics.

What Are *Visual Analogies*?

A visual analogy is a helpful way to learn new material based on what you already know from everyday life. It compares a chemical concept to something familiar. For example, the concept of electronegativity is compared to two men in a tug-of-war. Visualizing this helps you better understand the concept. Visual analogies accomplish several things:

1. They reduce your anxiety about learning the material and help you focus on the task at hand.

2. They force you to put an abstract concept into a familiar framework. This is referred to as *contextualized learning*.

3. They make the learning process more fun, relevant, and meaningful, so you can better understand and retain the information.

Whenever a visual analogy is used in this book, a small picture of it appears in the upper righthand corner of the module. This allows you to quickly reference a page visually, simply by flipping through the pages. As an alternative, the *Visual Analogy Index* that follows provides a table with all the visual analogies used and cross-references them to the appropriate page numbers.

Icons Used

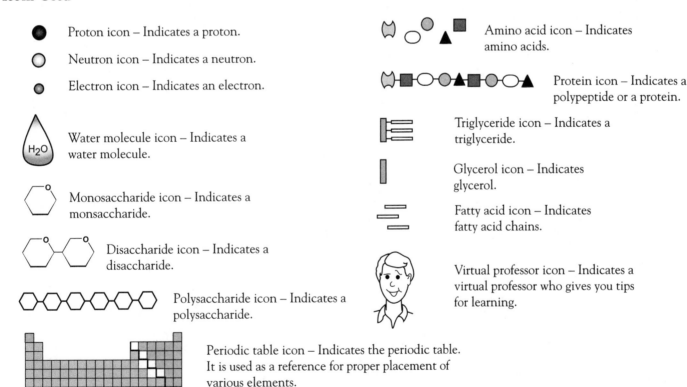

Proton icon – Indicates a proton.

Neutron icon – Indicates a neutron.

Electron icon – Indicates an electron.

Water molecule icon – Indicates a water molecule.

Monosaccharide icon – Indicates a monsaccharide.

Disaccharide icon – Indicates a disaccharide.

Polysaccharide icon – Indicates a polysaccharide.

Periodic table icon – Indicates the periodic table. It is used as a reference for proper placement of various elements.

Amino acid icon – Indicates amino acids.

Protein icon – Indicates a polypeptide or a protein.

Triglyceride icon – Indicates a triglyceride.

Glycerol icon – Indicates glycerol.

Fatty acid icon – Indicates fatty acid chains.

Virtual professor icon – Indicates a virtual professor who gives you tips for learning.

VISUAL ANALOGY INDEX

TOPIC	ANALOGY	ICON (S)	PAGE NO.
1. Atoms and elements— Electron-dot structures	A **compass** shows where to position the electrons		46, 47
2. Chemical bonds— Octet rule and chemical bonding	The symbol to represent the octet rule is an **eight ball**		54, 55
3. Chemical bonds— Octet rule and chemical bonding	A more stable product is like a **stable couple**		54
4. Chemical bonds— Covalent bond	A covalent bond is like **two people locking arms**		57, 161
5. Chemical bonds— Electronegativity	Electronegativity is like a **tug of war**		58, 59
6. Chemical bonds— Bond polarity	A dipole is like a **bar magnet**		59
7. Chemical equations— Understanding chemical equations	**Mixing ingredients together to make pancakes** is similar to the parts of a chemical equation		72
8. Chemical reactions— Combination reactions	Combination reactions are like **making a compound word**	"cupcake"	76
9. Chemical reactions— Decomposition reactions	Decomposition reactions are like **deconstructing a compound word**	"cup"	76
10. Chemical reactions— Replacement reactions	Replacement reactions are like **changing dance partners**		77, 120
11. Chemical reactions— Oxidation–reduction reactions	Redox reactions always occur in pairs like **two people playing catch**		79

VISUAL ANALOGY INDEX

VISUAL ANALOGY INDEX

TOPIC	ANALOGY	ICON (S)	PAGE NO.
22. Proteins—Functions of proteins	Transport proteins deliver cargo like **a car with a trailer**		164
23. Proteins—Functions of proteins	Hormones are chemical messengers that regulate body functions like a **carrier pigeon delivering a message**		165
24. Proteins—Functions of proteins	Catalysts such as enzymes are like **sparkplugs**		165
25. Proteins—Functions of proteins	Like **a shield** in battle, defense proteins protect the body against foreign pathogens		165
26. Proteins—Functions of proteins	Transmission proteins are like **cell phone towers** that send signals to your cell phone		165
27. Enzymes—Structure and function	The substrate fits in the active site on the enzyme like **a lock and key**		170, 171
28. Enzymes—Structure and function	Like a **baseball glove changes its shape to catch a baseball**, the active site changes its shape to bind the substrate		170, 171
29. Carbohydrates—Functions of carbohydrates	Like **gasoline for your car**, carbohydrates are used as energy for body cells		31, 79, 180, 181, 224
30. Carbohydrates—Functions of carbohydrates	Some carbohydrates serve as **markers** on cell surfaces		180, 181

VISUAL ANALOGY INDEX

TOPIC	ANALOGY	ICON (S)	PAGE NO.
31. Carbohydrates—Functions of carbohydrates	Stored carbohydrates are held in reserve like **items in a storage box**		180, 181
32. Nucleic acids—DNA replication	To replicate, DNA unzips like **a zipper**		194, 195
33. Nucleotides—RNA types	Messenger RNA is like a **copy of the portion of the DNA master blueprint**		196, 197, 200
34. Nucleotides—RNA types	Transfer RNA is like a **taxicab**		197
35. Nucleotides—RNA types	A ribosome is like a **protein factory**		197
36. Protein synthesis—The genetic code	The genetic code is like the **Morse code**		198, 199
37. Protein synthesis	The process of protein synthesis is like **making a new office building** from a master blueprint		196, 200
38. Lipids—Functions of lipids	Stored lipids are held in reserve like **items in a storage box**		206
39. Lipids—Functions of lipids	Lipids provide buoyancy to an organism like **a cork floating in water**		206

VISUAL ANALOGY INDEX

TOPIC	ANALOGY	ICON (S)	PAGE NO.
40. Lipids— Functions of lipids	Lipids act as **vehicles to transport their lipid cargo**		206
41. Lipids— Functions of lipids	Like a toy **sailboat can become a different boat by removing the sail**, a lipid precursor can be converted into another lipid		207
42. Lipids— Functions of lipids	Fat pads and cushions organs like **packing peanuts** protect items during shipment		207
43. Lipids— Functions of lipids	Lipids help your body retain heat like **attic insulation** helps a home retain heat		207
44. Lipids— Glycero- phospholipids	A cell membrane is like a **waterbed**		213
45. Bioenergetics— ATP hydrolysis	ATP hydrolysis is like a **good investment**		223
46. Bioenergetics— Carbohydrate metabolism— overview	Cellular respiration is like **the combustion of gasoline in a car engine**		31, 224
47. Bioenergetics— Glycolysis	Reduced coenzymes are like **a car with a trailer carrying an electron cargo**		218, 221, 225, 227, 229
48. Bioenergetics— Electron transport system	The electron transport system is like **passing a hot potato;** the hydrogen ion gradient is like **water behind the dam**		221, 231

Introduction to Chemistry

Description

Chemistry is the study of **matter,** which is anything that has mass or occupies space. As a practical example, let's imagine that you are a student in a classroom. First, we could turn inward and analyze the chemical nature of your body, which contains water, proteins, and lipids, to name just a few chemicals. Then we could examine your physical environment, such as the windows, walls, carpeting, chairs, and tables. Glass, metal, nylon, plastic, and wood may have been used to construct these items. Don't forget that the air in the room is composed of gases such as nitrogen, oxygen, and carbon dioxide.

To understand the importance of chemistry in more detail, let's consider the typical contents of a student's backpack. Here we find a wide variety of both natural and synthetic **chemicals** ranging from water, proteins, lipids, and carbohydrates to plastic, nylon, and vinyl. It's amazing to realize that everything within us and around us is made of chemicals.

Here is a handy table of the chemicals in the objects illustrated on the facing page (note: objects are listed alphabetically):

Object	Chemical Present
Antacid	Calcium carbonate or magnesium carbonate, aluminum (foil wrapping)
Apple	Water, carbohydrates, fiber, vitamins, antioxidants
Aspirin	Acetylsalicylic acid (ASA)
Backpack	Nylon (synthetic polymer), plastic (synthetic polymer), metal
Ham sandwich	Starch (natural polymer in bread), protein (natural polymer in ham), triglycerides (mayonnaise), salt (ham)
Hand sanitizer	Isopropyl alcohol, plastic (bottle)
Lip moisturizer	Petroleum jelly, menthol, vitamin E
Paper	Cellulose (natural polymer)
Pen	Plastic (in casing), ink (dyes in a solvent), metal (in the spherical point of the pen, often made of brass, which is an alloy of copper and zinc)
Pencil	Wood (soft woods such as incense cedar), graphite (form of carbon), rubber (synthetic polymer for eraser)
Sandwich bag	Plastic (synthetic polymer)
Soda pop	Water, carbohydrate (fructose), carbon dioxide gas, caffeine, plastic (bottle)
Textbook	Paper, cardboard (reinforced paper product), ink (dyes in a solvent), polyvinyl acetate or PVA (in glue for book binding)
Three-ring binder	Cardboard, vinyl, aluminum

Measurements

Description

Consider the importance of **measurement** to our daily lives. Whether it is the time needed to bake a cake, the height of our children, or the distance from our home to our destination, we are always measuring. Medical professionals measure blood pressures, drug dosages, body temperatures, and blood glucose levels. Scientists measure carbon dioxide levels in the atmosphere, and the magnitudes of earthquakes. In short, we all need to understand measurement.

A **number** and a **unit** are the two vital parts of any measurement. Each is useless without the other. For example, if someone asked you how long it takes to bake a cake and you answered "30," it would make no sense. Similarly, if you answered "minutes," that would be equally meaningless. The only acceptable answer contains *both* the number and the units. Therefore, the answer "30 minutes" makes perfect sense.

The **metric system** is the standardized system of measurement used by most countries around the world. It was developed by French scientists in 1790 and then spread to other countries. More recently, around 1960, the International System of Units (SI) was developed by scientific organizations and is called the modern metric system. It is similar to the metric system in some ways and different in others. Each system has its own advantages and disadvantages that will not be elaborated upon here. In this book we will use the metric system for consistency. Last, the reason that the United States uses a different system is that it originally adopted the English system of measurement based on miles, gallons, pounds, and the like.

Below is a table showing a comparison between the metric system and the SI system:

Measurement	Metric term (abbreviation)	SI term (abbreviation)
Length	Meter (m)	Meter (m)
Volume	Liter (L)	Cubic meter (m^3)
Mass	Gram (g)	Kilogram (kg)
Time	Second (s)	Second (s)
Temperature	Celsius (C)	Kelvin (K)

Let's examine the five different types of measurements given in the table above:

- **Length**—measures the distance between two points. The unit used by both the metric and SI systems is the **meter (m)**.

   100 centimeters (cm) = 1 meter (m)
   1,000 millimeters (mm) = 1 meter (m)

- **Volume**—measures the space occupied by a substance. The unit used by the metric system is the **liter**, and the SI system uses the **cubic meter (m^3)**

   1 milliliter (mL) = 1 cubic centimeter (cc or cm^3)
   1,000 milliliters (mL) = 1 liter (L)

- **Mass**—measures the amount of matter in an object. The unit used by the metric system is the **gram (g)**, and the SI uses the **kilogram (kg)**. It's important to note that mass is different from weight. Mass does not change with location, whereas weight does change because it's dependent on the force of gravity. For example, the weight of an astronaut on the moon is less than his or her weight on the earth because the force of gravity is less. Scientists report mass rather than weight.

   1,000 grams (g) = 1 kilogram (kg)

- **Time**—measures how long it takes an event to occur. The unit used by both the metric system and the SI system is the **second**.

   60 seconds = 1 minute
   60 minutes = 1 hour

- **Temperature**—measures the amount of heat in an object. The scale used by the metric system is the **Celsius (°C)** scale, and the SI system uses the **Kelvin (K)** scale. As a reference, let's examine the freezing and boiling point of water. On the Fahrenheit scale, water freezes at 32°F and boils at 212°F. On the Celsius scale, water freezes at 0°C and boils at 100°C. On the Kelvin scale, water freezes at 273 K and boils at 373 K.

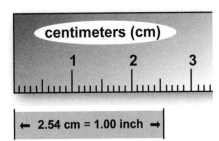

centimeters (cm)

1   2   3

← **2.54 cm = 1.00 inch** →

Length is measured in meters with a meterstick.

LENGTH

In a clinical setting, a hypodermic needle measures small volumes; in a lab setting, a graduated cylinder is used to measure large or small volumes.

VOLUME

A single serving carton of milk holds **237 mL = half a pint**

A MEASUREMENT
has two parts:
Number & Units

TEMPERATURE

MASS

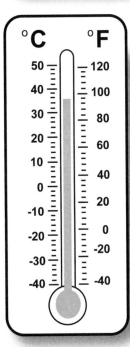

A thermometer is used to measure temperature.

Normal body temperature is **37.0 degrees C = 98.6 degrees F**

TIME

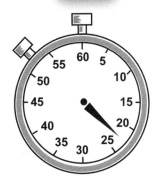

Time is measured in seconds as with a stopwatch.

Nerve impulses in the body are measured in milliseconds (ms).

1,000 ms = 1 second (s)

An electronic balance is used to measure mass.

The pen weighs **4.591 grams (g)** How much would a pack of ten pens weigh?

4.591 g

Description

Scientists deal with numbers that are often very large or very small. Writing out these numbers can be cumbersome because they contain lots of zeros. This led to the development of a helpful, shorthand method called **scientific notation**. For example, the total number of red blood cells in the body of an average adult is a whopping **5,000,000 cells**, and the length of a single red blood cell is a microscopic 0.000007 meters. Let's convert these two values to scientific notation:

$$5{,}000{,}000 \text{ cells} = 5 \times 10^6 \text{ cells}$$

$$0.000007 \text{ meters} = 7 \times 10^{-6} \text{ meters}$$

coefficient   power of ten

How to Convert Numbers to Scientific Notation

Notice that numbers in scientific notation have two parts: (1) a coefficient (≥ 1 and ≤ 9.99) and (2) a power of ten. The coefficient is always multiplied by the power of ten. In the first example given above, 5×10^6 **cells**, **5** is the coefficient, and 10^6 is the power of ten. How was the conversion made? We simply took the number, **5,000,000 cells**, and moved the decimal 6 places to the left until we arrived at a single–digit number other than zero. In this case, that number was 5, the coefficient. The number of places we moved the decimal gives us the exponent for the power of ten. In this example, it was **6**, which gave us 10^6. Any time the decimal is moved to the left, the power of ten is a *positive* number. Here are some other examples of **positive powers of ten**:

$$
\begin{aligned}
1{,}000{,}000 &= 1 \times 10^6 \\
100{,}000 &= 1 \times 10^5 \\
10{,}000 &= 1 \times 10^4 \\
1{,}000 &= 1 \times 10^3 \\
100 &= 1 \times 10^2 \\
10 &= 1 \times 10^1 \\
1 &= 1 \times 10^0
\end{aligned}
$$

Now let's look at the second example given above, 7×10^{-6} **meters** for the length of a single red blood cell. This time we took the number **0.000007 meters** and moved the decimal point to the right 6 places until we arrived at the number **7**, our coefficient. Because we moved the decimal **6** places, this determined our power of ten to be 10^{-6}. Notice that whenever the decimal point is moved to the right, the power of ten is a *negative* number. Here are some other examples of **negative powers of ten**:

$$
\begin{aligned}
0.1 &= 1 \times 10^{-1} \\
0.01 &= 1 \times 10^{-2} \\
0.001 &= 1 \times 10^{-3} \\
0.0001 &= 1 \times 10^{-4} \\
0.00001 &= 1 \times 10^{-5} \\
0.000001 &= 1 \times 10^{-6}
\end{aligned}
$$

Practice Problems

Identify the coefficient in each of the following:

1 4×10^{-3}

2 2.7×10^{4}

Convert the scientific notation into ordinary numbers:

3 1.2×10^{3}

4 6.2×10^{6}

5 8×10^{-4}

6 9.7×10^{-2}

Convert the following ordinary numbers to scientific notation:

7 535,000

8 0.009

9 0.04

10 6,331,000

11 0.0008

12 1,512,000,000

13 2,100

14 0.37

Answers

14 3.7×10^{-1}

9 4×10^{-2} 10 6,331 11 8×10^{-4} 12 1,512 $\times 10^{9}$ 13 2.1×10^{3}

1 4 2 2.7 3 1,200 4 6,200,000 5 0.0008 6 0.097 7 5.35×10^{5} 8 9×10^{-3}

Description

Scientists strive to be as correct as possible with each measurement they make. Unfortunately, measured numbers always have a degree of uncertainty because they are limited by the smallest unit of measure on the measuring tool. Let's say you were peeking through your kitchen window to check the temperature outdoors on an analog thermometer that uses a dial. You read the temperature as 43°F. But was it *exactly* 43°F? You can't be certain, because the smallest increment on your thermometer was 2 degrees. It might be 42.8°F or 43.1°F. In this example, **significant figures** are used to show certainty that the temperature is between 42°F and 44°F but there is some uncertainty in the last digit of this measurement. Our measurement of 43°F has two significant figures in it.

To determine the significant figures for any number, we need to use the following rules.

Rules for Determining the Number of Significant Figures (digits)

A number **IS** a significant figure if it is ...

1 **A nonzero number** (1, 2, 3, 4, 5, 6, 7, 8, 9)

Example: 47.6 g has **3** significant figures; 0.5216 m has **4** significant figures

2 **A zero between nonzero numbers**

Example: 8.02 L has **3** significant figures; 0.1706 m has **4** significant figures

3 **A zero at the end of a number with a decimal point**

Example: 5.00 mL has **3** significant figures; 9.730 mm has **4** significant figures

4 **The coefficient of a number written in scientific notation**

Example: $\mathbf{5.0} \times 10^6$ m has **2** significant figures; $\mathbf{8.40} \times 10^{-3}$ g has **3** significant figures

           coefficient   coefficient

A number is **NOT** a significant figure if it is ...

1 **A zero before or after the decimal**

Example: 0.4 has **1** significant figure; 0.003 m has **1** significant figure; 0.079 g has **2** significant figures;

2 **A zero at the end of a large number without a decimal point**

Example: 260,000,000 L has **2** significant figures; 3,456,000 m has **4** significant figures

Practice Problems

Determine the number of **significant figures** (digits) in each of the following numbers and write it in the blank:

1 86.2 __________

2 20.06 __________

3 7,000,000 __________

4 0.0093 __________

5 4.1×10^{12} __________

6 8.60×10^{-3} __________

7 32.04 __________

8 1,186.0 __________

9 27.00 __________

10 407.06 __________

Answers

1 3 2 4 3 1 4 2 5 2 6 3 7 4 8 5 9 4 10 5

Description

Prefixes are commonly used in metric and SI measurement systems, so they have to be learned. For example, a woman's height is measured in *centi*meters, her mass in *kilo*grams, and her hourly production of saliva in *milli*liters. Common words in the English language are derived from these prefixes and help you decode their meaning: *cents* in a dollar (*centi* = 100), *decade* (*deca* = 10), and *kilo*meter (*kilo* = 1,000).

The tables below list common prefixes:

Prefixes that INCREASE ↑ the size of the unit

Prefix	Symbol	Meaning (*in words*)	Meaning (*in numbers*)	Power of Ten
Tera–	T	trillion	1,000,000,000,000	10^{12}
Giga–	G	billion	1,000,000,000	10^{9}
Mega–	M	million	1,000,000	10^{6}
Kilo–	k	thousand	1,000	10^{3}
Hecto–	h	hundred	100	10^{2}
Deca–	da	ten	10	10^{1}

Prefixes that DECREASE ↓ the size of the unit

Prefix	Symbol	Meaning (*in words*)	Meaning (*in numbers*)	Power of Ten
Deci–	d	tenth	0.1	10^{-1}
Centi–	c	hundredth	0.01	10^{-2}
Milli–	m	thousandth	0.001	10^{-3}
Micro–	μ	millionth	0.000 001	10^{-6}
Nano–	n	billionth	0.000 000 001	10^{-9}
Pico–	p	trillionth	0.000 000 000 001	10^{-12}

Practice Problems

Fill in the blanks with the correct **prefix:**

Example: **0.02 meter = 2 centi meters**

1 0.01 meter = 1 __________ meter

2 0.03 meter = 3 __________ meters

3 0.001 liter = 1 __________ liter

4 0.006 liter = 6 __________ liters

5 1,000 grams = 1 __________ gram

6 2,000 grams = 2 __________ grams

7 10 meters = 1 __________ meter

8 30 meters = 3 __________ meters

Fill in the blanks with the correct **power of 10:**

Example: **0.05 liter = 5×10^{-2} liter**

9 0.03 meter = $3 \times$ __________ meter

10 0.001 liter = $1 \times$ __________ liter

11 0.006 liter = $6 \times$ __________ liter

12 100 grams = $1 \times$ __________ grams

13 2,000 grams = $2 \times$ __________ grams

14 10 meters = $1 \times$ __________ meters

Answers

1 centi- 2 centi- 3 milli- 4 milli- 5 kilo- 6 kilo- 7 deca-

8 deca- 9 10^{-2} 10 10^{-3} 11 10^{-3} 12 10^{2} 13 10^{3} 14 10^{1}

Description

What determines whether a metal coin sinks or a piece of wood floats in water? **Density**. Anything that has a density less than water floats; anything that has a density greater than water sinks. You might think that the reason the metal coin sinks is that it has a greater mass, which makes it "heavier," but you would be only half right because density is determined by the mass divided by the volume of an object. By calculating this physical property, we get an idea how tightly packed the atoms are within the object.

Density (D) is equal to the **mass (M)** of an object divided by its **volume (V)**:

$$D = M/V$$

The table below gives the approximate densities of some common substances.

DENSITIES of some COMMON SUBSTANCES

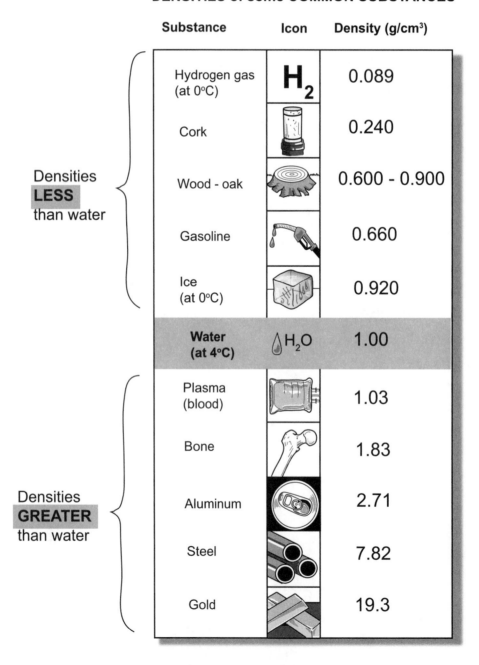

Substance	Icon	Density (g/cm³)
Hydrogen gas (at 0°C)	H_2	0.089
Cork		0.240
Wood - oak		0.600 - 0.900
Gasoline		0.660
Ice (at 0°C)		0.920
Water (at 4°C)	H_2O	1.00
Plasma (blood)		1.03
Bone		1.83
Aluminum		2.71
Steel		7.82
Gold		19.3

Densities **LESS** than water

Densities **GREATER** than water

Practice Problems

1 Three different objects are placed in a tub of water. Based on the illustration below, select the letter for the object that has the *lowest* density. Explain.

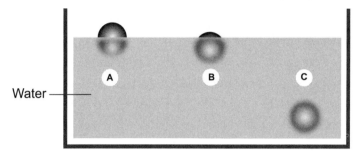

Water

2 You want to determine the density of a silver coin. First you place it on a balance to determine that it has a mass of 15.8 grams. Then you place it in a graduated cylinder filled with water. After dropping the coin in the water, it raises the water level by 1.50 mL (cm³). Calculate the density and show your calculation.

3 Calculate the density for each of the following (be sure to show the *units* for each):

a An object that has a mass of 75.6 g and a volume of 53.4 cm³.

b An object that has a mass of 0.251 g and a volume of 0.399 mL.

c An object that has a mass of 167 g and a volume of 85.8 mL.

Answers

c 167 g/85.8 mL = 1.95 g/mL

b 0.251 g/0.399 mL = .629 g/mL

3 a 75.6 g/53.4 cm³ = 1.42 g/cm³

2 15.8 g/1.50 mL = 10.5 g/mL

1 Object A has the lowest density as it floats the highest in the tub of water.

Matter
and
Energy

Description

Matter can be found in three different states: **solid**, **liquid**, or **gas**. For example, water is a liquid coming out of your kitchen faucet, but it can also be in the form of a solid (ice cubes) or a gas (steam). The properties of each of these states is dependent on factors such as temperature and pressure. Let's examine each of these in more detail.

Solid

The illustration on the facing page shows three examples of solids: copper pipes, ice cubes, and lard. Look around you to see many more examples. Solids have a firm, rigid structure, so it makes sense that the particles that compose them are packed closely together in an orderly arrangement. In other words, what we observe with the naked eye is reflected in their molecular structure. Here is a summary of the properties of a **solid**:

- **Shape:** has a defined shape
- **Volume:** has a defined volume (*at a given temperature*)
- **Arrangement:** particles have orderly arrangement; compact, close together
- **Movement:** particles move very slowly

Liquid

Depicted on the facing page are three liquids: water, liquid mercury, and gasoline. While not as rigid as solids, liquids are dynamic because they are able to move and flow like water in a stream or plasma in your blood. This capacity to flow is reflected in the more random arrangement of their particles at the molecular level. Here is a summary of the properties of a **liquid**:

- **Shape:** takes the shape of its container (*at a given pressure*)
- **Volume:** has a defined volume (*at a given temperature*)
- **Arrangement:** particles have random arrangement; close together but not as close as solids
- **Movement:** particles move at a medium speed

Gas

We see three examples of gases on the facing page: balloons filled with helium gas, a cylinder of oxygen gas, and a teapot that has water vapor escaping from it. Only the water vapor is visible, while the gases in the other examples are invisible. But we know indirectly that the gases are present. When a balloon is filled with gas, we can observe it enlarge and become a spherical shape. The force exerted by the fast-moving gas particles keeps the balloons inflated. Gases are the lightest and least dense, so it's easy to predict that the particles that compose them are the farthest apart. Here is a summary of the properties of a **gas**:

- **Shape:** takes the shape of its container
- **Volume:** fills the space in its container (*at a given temperature*)
- **Arrangement:** particles are scattered; far apart
- **Movement:** particles move fast

Solid

In a solid, particles are packed closely together and arranged in a very organized manner

particle

Copper pipes

Ice cubes

Lard

Liquid

Particles in a liquid are close together but have a more randomized arrangement

particle

Water

Liquid mercury

Gasoline

Gas

Particles in a gas are far apart from each other

particle

Balloons filled with helium gas

Cylinder filled with oxygen gas

Water vapor escaping from a teapot

27

Description

Whenever matter converts from one state to another, such as a solid to a liquid or a gas to a liquid, it undergoes what is called a *change of state*. These changes occur when heat is either added or removed from matter.

Gas and Liquid States

A gas can be converted to the liquid state, or a liquid can be converted to the gas state. Here are the terms for each specific change of state, along with an example of each process:

- **Condensation**—changing states from a gas to a liquid through the removal of heat

  Example: Water droplets form on the outside of a glass of lemonade on a hot, humid summer day. As the hot, moist air comes in contact with the cold surface of the glass, the water vapor in the air (gas) loses heat to the environment as it transforms into a liquid.

- **Vaporization**—changing states from a liquid to a gas through the addition of heat

  Example: Water heated in a tea kettle is converted into steam. The addition of heat from the stove causes the liquid water to reach its boiling point, leading to the conversion into a gas.

Liquid and Solid States

A liquid can change into a solid state, or a solid can change into a liquid state. Here are the terms for each specific change of state, along with an example of each process:

- **Freezing**—changing from a liquid state to a solid state through the removal of heat

  Example: Liquid water placed in ice cube trays and put in a freezer will form solid ice cubes. The freezer removes heat from water, so it becomes ice and maintains a colder temperature at or below water's freezing point. This leads to the change of state.

- **Melting**—changing from a solid state to a liquid state through the addition of heat

  Example: Ice cubes float in a glass of water on a hot day but are gradually converted into liquid water. The heat from the environment causes the ice cubes to melt.

Solid and Gas States

A solid can be converted into a gas state, or a gas can be converted into a solid state. Here are the terms for each specific change of state, along with an example of each process:

- **Deposition**—changing from a gas state directly into a solid state through the removal of heat

  Example: The process of frost formation. When cloud cover is lost on a cold winter night, heat is released into the atmosphere. This causes moisture in the cold air to rapidly condense and quickly deposit in the form of frost on various surfaces such as windows, cars, trees, plants, and roads.

- **Sublimation**—changing from a solid state, through addition of heat, without forming a liquid first

  Example: The process of creating freeze-dried foods. Consider the freeze-dried strawberries in your breakfast cereal. First the strawberries are frozen, then placed in a vacuum chamber, where the fruit loses its moisture and shrivels up as the ice sublimes. This is a great way to preserve foods at room temperature, because they lack the moisture needed by bacteria to grow, thereby preventing spoilage.

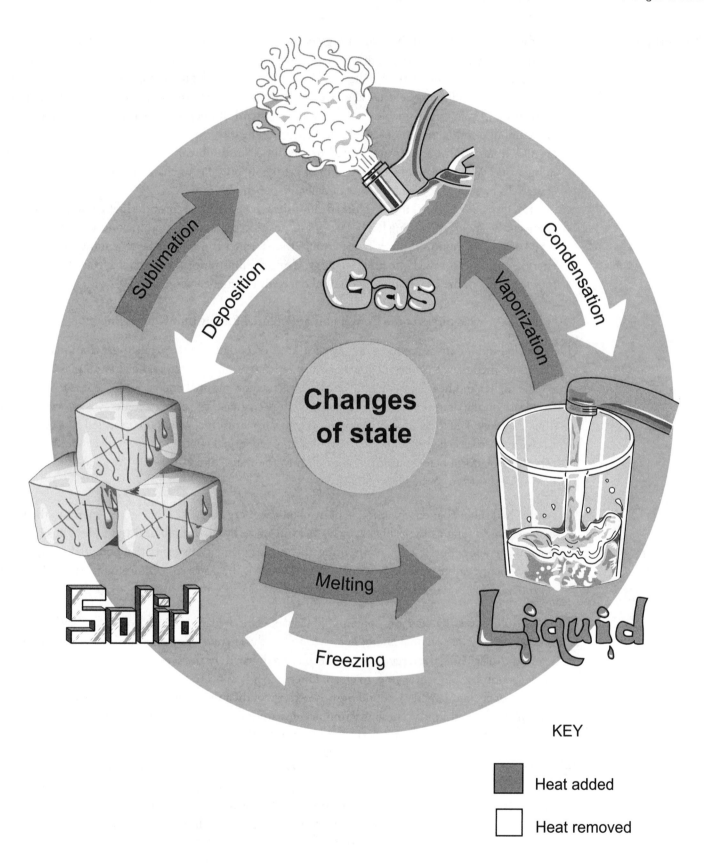

KEY

Heat added

Heat removed

Description

We casually throw around the term **energy** without understanding what it really means. Let's put it in an everyday context. Imagine that you are working on your laptop at the library, learning about energy, the ability to do work. The lighting in the library, your laptop computer, and the cell phone you are charging all require **electrical energy** to function. Your hands would be unable to type on the keyboard and your brain unable to think without **chemical energy** to fuel muscle movement and nerve impulse conduction, respectively. Suddenly, you hear the loud "ping" sound from a microwave oven. A library staff member just heated up her lunch in the break room—another form of energy doing work. Whether it's turning on a light or getting up out of your chair, all these processes require energy. To be sure, all of the processes of daily life depend on it.

Energy is best divided into two broad categories: **potential energy** and **kinetic energy**. Potential energy is the energy something has because of its position or stored energy. In contrast, kinetic energy is the energy of motion. Energy is commonly transformed from one type to another, so potential energy can be transformed into kinetic energy. Using the illustrations on the facing page, let's examine four examples of the conversion of potential energy into kinetic energy.

Examples of the Conversion of Potential Energy into Kinetic Energy

1. Water behind a dam is a good example of a kind of potential energy called **gravitational energy**. This is a function of the force of gravity and is dependent on two factors, height and mass. The rule is: The higher and heavier an object, the greater is the gravitational energy, and vice versa. When the floodgate on the dam opens, water flows through the dam so the potential energy is converted into kinetic energy. The force of the flowing water can turn a turbine, which is connected to a generator, to create hydroelectric power. Other examples of gravitational energy include a diver standing at the top of his platform before launching himself into the water, or a boulder at the edge of a steep cliff before it rolls down to the bottom.

2. A stretched rubber band is an example of **mechanical energy** because it has stored energy in the form of tension. As long as the rubber band is held in place by the thumb and the index finger, it has potential energy. But the act of shooting the rubber band transforms the potential energy into kinetic energy. Another example of mechanical energy is the act of squeezing a metal spring between both hands before releasing it to allow it to recoil.

3. The gasoline used by your car contains molecules composed of both hydrogen and carbon called hydrocarbons. The potential energy here is found in the chemical bonds of the hydrocarbons, so it is referred to as **chemical energy**. These molecules are burned, in the presence of oxygen, in a combustion reaction by your car engine, to create kinetic energy and move the wheels of your car. Another example of chemical energy is found in the nutrients of the foods that you consume, such as proteins, carbohydrates, and lipids. These macromolecules are used by your cells to fuel all cellular activities such as muscle contraction and nerve impulse conduction.

4. A battery stores another type of potential energy—electrical energy. When the batteries are placed in a flashlight and the switch is turned on to complete the electrical circuit, electrons flow through the battery from the negative end to the positive end. This is the kinetic energy used to light the bulb of the flashlight. Anything powered by a battery, such as a cell phone, a computer, or a garage door opener are additional examples of electrical energy.

POTENTIAL ENERGY	is converted to	KINETIC ENERGY

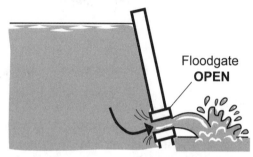

Floodgate **CLOSED**

Floodgate **OPEN**

① Because of its higher position, the **water behind the dam** has potential energy.

① As the floodgate opens, water moves through the dam, converting potential energy into kinetic energy, which, in turn, can be used to turn a turbine to generate hydroelectric power.

② A **stretched rubber band** - has stored energy because of its high degree of tension.

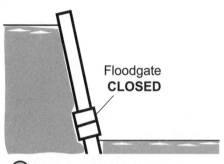

② The act of shooting the rubber band converts potential energy into kinetic energy.

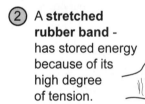

③ **Gasoline** has stored energy in the chemical bonds of the gasoline.

③ Gasoline is burned in a combustion reaction in a car engine to turn the wheels of the car.

CAR   HEAT

COMBUSTION REACTION

energy to turn wheels

④ A **battery** stores electrical energy.

④ The energy from the battery can be used to generate electricity for a flashlight.

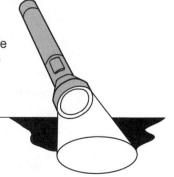

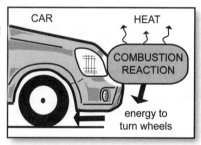

Conservation of Energy

First we must understand what is meant by a "closed system." In a closed system, such as a sealed container, energy in the form of heat can be exchanged with its surroundings but matter cannot be exchanged. A greenhouse is a good example of this. Heat can be added or removed from the greenhouse. On a hot, sunny day, heat is added, raising the temperature inside. In the evening, heat is removed, lowering the temperature. Conversely, the matter inside the greenhouse, such as the soil, rocks, and plants, cannot leave. Similarly, the cars, buildings, and other matter outside the greenhouse can't enter. Knowing this, now we can better understand the law for the **conservation of energy**:

> In a closed system, the total amount of energy remains constant; energy is neither created nor destroyed, it is only *transformed* from one state to another.

Using the example of the diver, the illustration on the facing page shows potential energy being transformed into kinetic energy. Here is a summary of each step shown:

1. After climbing to the top of the platform, the diver has a high degree of potential energy because of his elevated position above ground level. He has not jumped off the diving platform yet, he has no kinetic energy or energy of motion. The graph indicates that he has 100% potential energy and 0% kinetic energy.

2. The moment the diver jumps off the platform, he begins to transform some of his potential energy into kinetic energy. As he falls toward the water, his kinetic energy increases as his potential energy decreases by the same proportion. The graph shows that at this point in his dive, his potential energy is about 60% while the remaining 40% has been converted into kinetic energy.

3. The moment the diver hits the water, he has returned to ground level, having converted all his potential energy to kinetic energy. In other words, at this time, he has 100% kinetic energy and 0% potential energy.

During the dive, potential energy was simply transformed into kinetic energy. No energy was created or destroyed.

Conservation of Mass

This is the law for the **conservation of mass**:

> In chemical reactions, the MASS of the reactants = the MASS of the products.

Imagine you weigh three items from your kitchen—flour, eggs, and milk. Then you combine them in a bowl to make something new—pancake batter. Common sense tells you that the mass of the batter would equal the sum of the three ingredients. This is stated by the law of the conservation of mass. The same is true for chemical reactions. Consider the example of the nail changing into a rusted nail on the facing page. This is a chemical reaction between iron and oxygen to form a new compound, iron oxide. After sitting outside for a long time, the iron in the nail reacts with the oxygen in the atmosphere to form rust—iron oxide. Let's say that we originally weighed the new nail and discovered that it was 40.6 g. Then we later weighed the rusted nail and determined that it weighed 75.5 g. The difference between these two figures (75.5 g − 40.6 g) gives us the mass of the oxygen, 34.9 g. Notice that no mass was created or destroyed during the chemical reaction. Instead, the mass of the reactants equals the mass of the final product.

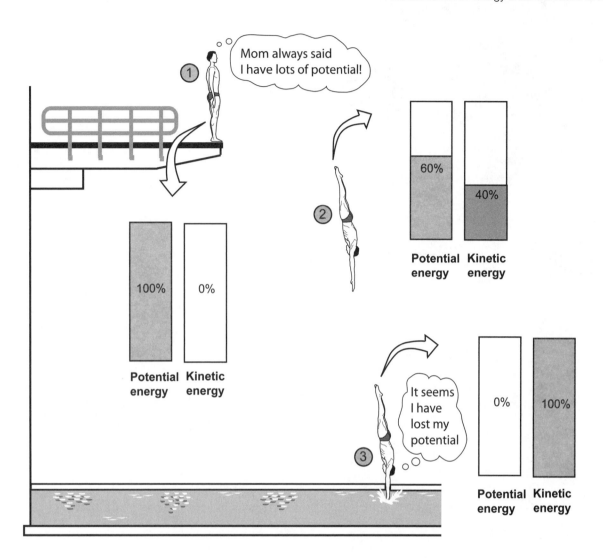

Example:	A **nail** becomes a **rusted nail**
Chemical reaction:	**Iron** (in the nail) reacts with **oxygen** (from the environment) to become **iron oxide** (rust)
Chemical equation:	$4Fe$ (iron) $+$ $3O_2$ (oxygen) $\longrightarrow$ $2Fe_2O_3$ (iron oxide)
Visual:	+ O_2 $\longrightarrow$
Mass:	40.6 g $+$ 34.9 g $=$ 75.5 g

Atoms
and
Elements

Description

The **periodic table of elements** (periodic table) lists all the elements, both natural and man-made, and organizes them by grouping similar elements together. In fact, the term *periodic* refers to trends in the properties of the elements. The Russian chemist Dmitri Mendeleev presented his periodic table for the first time in 1869. The original table has evolved over the years as new elements have been discovered and added to it. By classifying the elements to show trends, the table did not come out to look like a perfect square or rectangle, but is an asymmetrical shape. Rather than examining the periodic table as a list of things to memorize, let's look instead at it as a tool to help us understand common relationships between different elements.

Metals, Nonmetals, and Metalloids

The periodic table on the facing page color-codes the elements into metals, nonmetals, and metalloids:

- **Metals**—most elements, such as copper (Cu), iron (Fe), lead (Pb), and gold (Au). They tend to be shiny solids that are pliable and good conductors of electricity.

- **Nonmetals**—elements such as hydrogen (H), oxygen (O), and helium (He). Some are solids; others are liquids, or gases. All nonmetals are poor conductors of electricity.

- **Metalloids**—consists of six elements: boron (B), silicon (Si), germanium (Ge), arsenic (As), antimony (Sb), and tellurium (Te). Notice they form a boundary between metals and nonmetals. In this sense, they are the blending of metals and nonmetals because they contain properties of both. Notice the thicker lines forming a staircase that runs through the metalloids. This staircase separates metals on the left from nonmetals on the right.

Periods and Groups

Examine the periodic table on the facing page to differentiate between a period and a column.

- **Period**—any *horizontal row* on the periodic table. The 7 different periods are numbered 1–7.
  - *Example:* Period 1 contains only 2 elements beginning with hydrogen (H) and ending with helium (He)
  - *Example:* Period 5 contains 18 elements beginning with rubidium (Rb) and ending with xenon (Xe)

- **Group** (*family*)—any *vertical column* on the periodic table. The 18 different groups are labeled in two different ways. The U.S. system uses 1A–8A and 1B–8B. The newer International Union of Pure and Applied Chemistry (IUPAC) system simply numbers them 1–18. Both systems are shown because both are used.
  - *Example:* Group 2A (2) contains 6 elements beginning with beryllium (Be) and ending with radium (Ra)
  - *Example:* Group 4A (14) contains 6 elements beginning with carbon (C) and ending with ununquadium (Uuq)
  - *Example:* Group 4B (4) contains 6 elements beginning with titanium (Ti) and ending with rutherfordium (Rf)

Common Group Names

It's helpful to familiarize yourself with the following five common group names:

- **Alkali metals**—group 1A (1): soft metals such as sodium (Na) and potassium (K) that have 1 electron in their valence shell and are highly reactive, especially with halogens. All are particularly reactive with water.

- **Alkaline earth metals**—group 2A (2): soft, often silver-colored metals such as magnesium (Mg) and calcium (Ca) that have 2 electrons in their valence shell and are also very reactive, albeit less so than the alkali metals.

- **Transition elements**—group 1B-8B (3–12): also called *transition metals*; all are very hard with high melting and boiling points and are excellent conductors of electricity.

- **Halogens**—group 7A (17): nonmetals such as fluorine (Fl) and chlorine (Cl) that have 7 electrons in their outer shell and are unusually reactive with alkali metals and alkaline earth metals.

- **Noble gases**—group 8A (18): elements such as helium (He) and neon (Ne) that are all gases at room temperature. They have a full outer shell of electrons, so all are quite nonreactive and have low boiling points.

PERIODIC TABLE of ELEMENTS

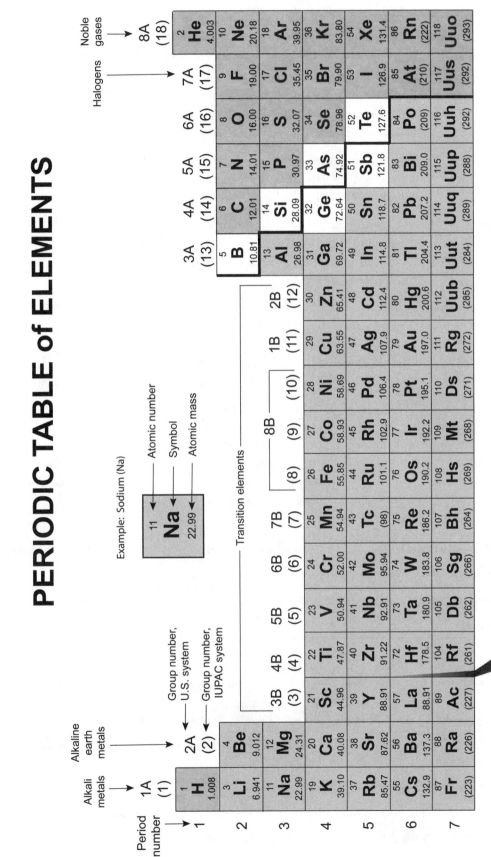

Element Names

The table on the facing page gives an alphabetical listing of all the elements and their symbols. First, let's consider the naming process, then the symbols. Elements are named after a variety of things such as scientists, mythical characters, places, and descriptions. Here are a few examples of each:

- **Scientists**
  Einsteinium—after the physicist Albert Einstein
  Mendelevium—after the chemist Dmitri Mendeleev
  Lawrencium—after the physicist Ernest O. Lawrence

- **Mythical characters**
  Promethium—after Prometheus, the powerful deity from Greek mythology
  Thorium—after Thor, the Norse god of thunder
  Titanium—from the titans of Greek mythology

- **Places**
  reveal where the element was discovered:
  Americium—after the Americas
  Californium—after the state of California
  Europium—after Europe

- **Descriptions**
  reveal some characteristic of the element:
  Bismuth—from the German word *Wismuth*, meaning "white mass" based on its appearance
  Carbon—from the Latin word *carbo*, which means "charcoal"
  Radium—from the Latin word *radius*, meaning "ray" because of its radioactivity

Symbols

Symbols are one-, two-, or three-letter abbreviations for the element names that follow these rules: (1) first letter is always capitalized, (2) other letters are lowercase. The one-letter symbols are typically taken from the first letter of the element name, and the two-letter symbols are often (but not always) taken from the first two letters of the element name. Three-letter symbols are commonly three key letters within the element name. Here are some examples (*Note:* bold-faced letter(s) in the element name indicate the origin of the symbol):

One-Letter Symbols	Two-Letter Symbols	Three-Letter Symbols
H hydrogen	**He** helium	**Uub** ununbrium
O oxygen	**Li** lithium	**Uup** ununpentium
N nitrogen	**Ne** neon	**Uus** ununseptium

In some cases, the symbols are not taken from letters in the element name but, rather, reflect the Greek or Latin origin of the word. Here are some examples:

- **Na** for **sodium** comes from the Latin word *natrium*, meaning "soda."
- **Pb** for **lead** comes from the Latin word *plumbum*, which is where we get the English word "plumbing."
- **Fe** for **iron** is derived from the Latin word *ferrum*, meaning "iron."
- **Au** for **gold** is taken from the Latin word *aureum*, meaning "glow of sunrise."
- **K** for **potassium** is derived from the Latin word *kalium*, meaning "potash."

Understanding the element names and symbols makes your study of chemistry much easier.

	Name	Symbol	Atomic #
A	Actinium	Ac	89
	Aluminum	Al	13
	Americium	Am	95
	Antimony	Sb	51
	Argon	Ar	18
	Arsenic	As	33
	Astatine	At	85
B	Barium	Ba	56
	Berkelium	Bk	97
	Beryllium	Be	4
	Bismuth	Bi	83
	Bohrium	Bh	107
	Boron	B	5
	Bromine	Br	35
C	Cadmium	Cd	48
	Calcium	Ca	20
	Californium	Cf	98
	Carbon	C	6
	Cerium	Ce	58
	Cesium	Cs	55
	Chlorine	Cl	17
	Chromium	Cr	24
	Cobalt	Co	27
	Copper	Cu	29
	Curium	Cm	96
D	Darmstadtium	Ds	110
	Dubnium	Db	105
	Dysprosium	Dy	66
E	Einsteinium	Es	99
	Erbium	Er	68
	Europium	Eu	63
F	Fermium	Fm	100
	Fluorine	F	9
	Francium	Fr	87
G	Gadolinium	Gd	64
	Gallium	Ga	31
	Germanium	Ge	32
	Gold	Au	79
H	Hafnium	Hf	72
	Hassium	Hs	108

	Name	Symbol	Atomic #
	Helium	He	2
	Holmium	Ho	67
	Hydrogen	H	1
I	Indium	In	49
	Iodine	I	53
	Iridium	Ir	77
	Iron	Fe	26
K	Krypton	Kr	36
L	Lanthanum	La	57
	Lawrencium	Lr	103
	Lead	Pb	82
	Lithium	Li	3
	Lutetium	Lu	71
M	Magnesium	Mg	12
	Manganese	Mn	25
	Meitnerium	Mt	109
	Mendelevium	Md	101
	Mercury	Hg	80
	Molybdenum	Mo	42
N	Neodymium	Nd	60
	Neon	Ne	10
	Neptunium	Np	93
	Nickel	Ni	28
	Niobium	Nb	41
	Nitrogen	N	7
	Nobelium	No	102
O	Osmium	Os	76
	Oxygen	O	8
P	Palladium	Pd	46
	Phosphorus	P	15
	Platinum	Pt	78
	Plutonium	Pu	94
	Polonium	Po	84
	Potassium	K	19
	Praseodymium	Pr	59
	Promethium	Pm	61
	Protactinium	Pa	91
R	Radium	Ra	88
	Radon	Rn	86
	Rhenium	Re	75

	Name	Symbol	Atomic #
	Rhodium	Rh	45
	Roentgenium	Rg	111
	Rubidium	Rb	37
	Ruthenium	Ru	44
	Rutherfordium	Rf	104
S	Samarium	Sm	62
	Scandium	Sc	21
	Seaborgium	Sg	106
	Selenium	Se	34
	Silicon	Si	14
	Silver	Ag	47
	Sodium	Na	11
	Strontium	Sr	38
	Sulfur	S	16
T	Tantalum	Ta	73
	Technetium	Tc	43
	Tellurium	Te	52
	Terbium	Tb	65
	Thallium	Tl	81
	Thorium	Th	90
	Thulium	Tm	69
	Tin	Sn	50
	Titanium	Ti	22
	Tungsten	W	74
U	Ununbium	Uub	112
	Ununhexium	Uuh	116
	Ununoctium	Uuo	118
	Ununpentium	Uup	115
	Ununquadium	Uuq	114
	Ununseptium	Uus	117
	Ununtrium	Uut	113
	Uranium	U	92
V	Vanadium	V	23
X	Xenon	Xe	54
Y	Ytterbium	Yb	70
	Yttrium	Y	39
Z	Zinc	Zn	30
	Zirconium	Zr	40

Atomic Structure

The illustration on the facing page shows a cylinder of **helium** gas to represent the element helium and graphite in a pencil to represent the element **carbon**. Helium gas is used to fill balloons and we all know how pencils are used. Each of these elements, like all others, is made of many identical **atoms**. With the exception of hydrogen, each atom contains three subatomic particles: **protons, neutrons,** and **electrons**. The protons and neutrons are packed tightly together at the center of the atom called the **nucleus**. Electrons are particles found moving at high speed in a spherical region around the nucleus called the **electron cloud**. Here is a summary of the properties of each subatomic particle:

- **Proton**—a particle that has a **positive** (+) charge (remember the phrase "Protons are Positive")

- **Neutron**—a particle that is **neutral** (no charge). (remember the phrase "Neutrons are Neutral")

- **Electron**—a small, almost mass-less particle with a **negative** charge

Atomic Number

Atomic number = number of protons in an atom

Every element on the periodic table has a different atomic number, which indicates the number of protons in a single atom of that element. The illustrations show helium as atomic number 2, telling us that it has only 2 protons (+2); carbon, with atomic number 6, has 6 protons (+6). The net charge on any atom is always zero, which means that the number of protons and electrons is always equal. Notice that the helium atom has 1 proton and 1 electron and the carbon atom has 6 protons and 6 electrons. Because protons are always positive and electrons are always negative, the net charge of both atoms is zero.

Practice Problems

Determine the number of protons in an atom of the following elements:

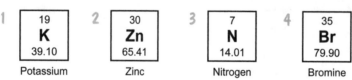

| 1 | 19
K
39.10
Potassium | 2 | 30
Zn
65.41
Zinc | 3 | 7
N
14.01
Nitrogen | 4 | 35
Br
79.90
Bromine |

5 Are the total number of protons and electrons in an atom the same or different? Explain.

6 Which subatomic particle has a no charge?

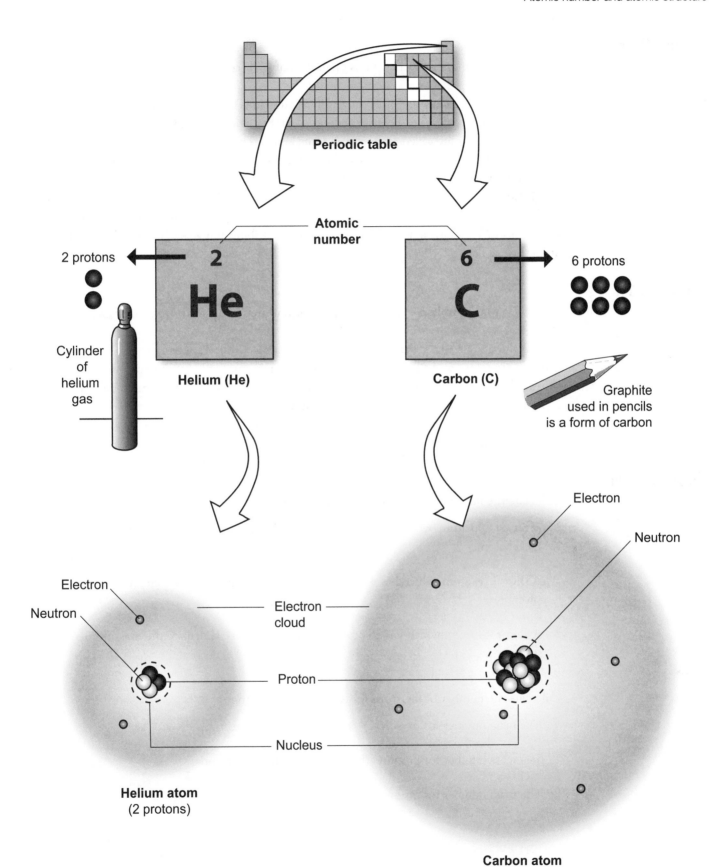

Periodic table

Atomic number

2 protons

2

He

Cylinder of helium gas

Helium (He)

6 protons

6

C

Carbon (C)

Graphite used in pencils is a form of carbon

Electron

Neutron

Electron

Neutron

Electron cloud

Proton

Nucleus

Helium atom
(2 protons)

Carbon atom
(6 protons)

Atomic Mass

> Atomic mass = average mass of all the naturally occurring isotopes of that element

On the periodic table (see p. 37), the atomic mass number is found below the symbol of the element. For example, on the facing page, helium (He) has an atomic mass of **4.003**. The units used for measuring very small particles are called atomic mass units (amu). These represent the atomic mass expressed in grams (g). Why isn't the atomic mass a whole number? It is calculated as an *average* of all the naturally occurring **isotopes** (see p. 50) of that element. By comparison, the atomic mass for carbon is given as **12.01**. Notice that it also is not a whole number for the reason already mentioned. The atomic mass tells us the mass of the nucleus of an atom, which contains both protons and neutrons. Why not consider the electrons? They have a negligible mass, so they contribute almost nothing to the overall mass of the atom.

Mass Number

> Mass number = number of **protons and neutrons** in the nucleus of an atom

Unlike the atomic mass, the mass number is always a whole number because it tells us the total number of protons and neutrons in the nucleus of an atom. Let's examine the mass number for each of the illustrated atoms. As the second element on the periodic table, the helium atom has only 2 protons and 2 neutrons. By adding these numbers together (2 + 2) we determine that the mass number is 4. In comparison, the carbon atom has 6 protons and 6 neutrons, giving it a mass number of 12 (by adding 6 + 6).

Practice Problems

Calculate the **mass number** for each of the following:

1 A nickel (Ni) atom, which has 28 protons and 31 neutrons. Calculate the **mass number** for this nickel atom.

2 A silver (Ag) atom, which has 47 protons and 61 neutrons. Calculate the **mass number** for this silver atom.

3 A sulfur (S) atom, which has 16 protons and 16 neutrons. Calculate the **mass number** for this sulfur atom.

Calculate the number of **neutrons** for each of the following:

4 The mass number for oxygen (O) is 16 and it has 8 protons.

5 The mass number for bromine (Br) is 80 and it has 35 protons.

6 The mass number for magnesium (Mg) is 24 and it has 12 protons.

Answers

1 59 2 108 3 32 4 8 5 45 6 12

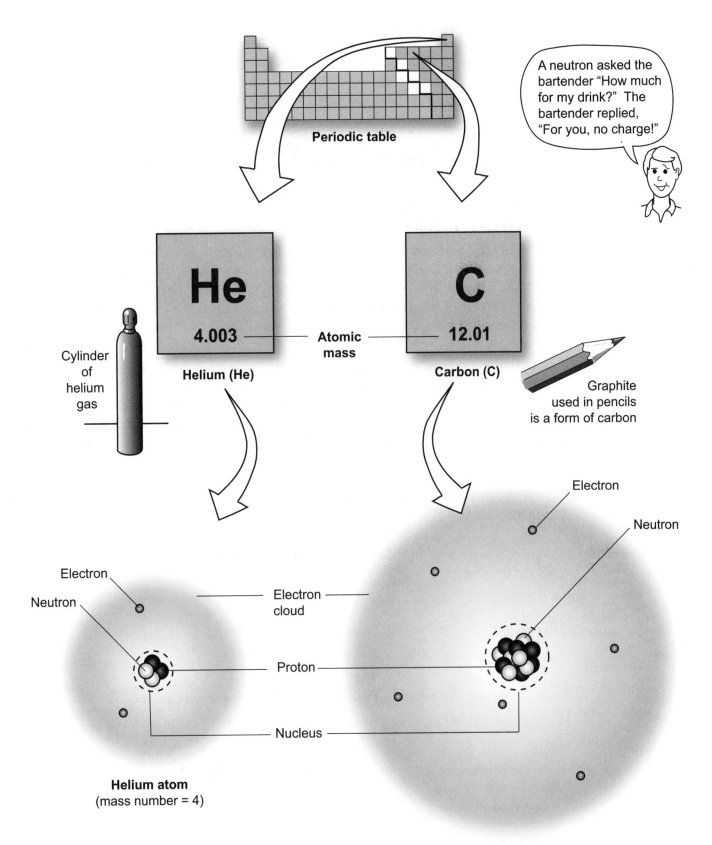

Periodic table

A neutron asked the bartender "How much for my drink?" The bartender replied, "For you, no charge!"

Cylinder of helium gas

He
4.003
Helium (He)

Atomic mass

C
12.01
Carbon (C)

Graphite used in pencils is a form of carbon

Electron

Neutron

Electron cloud

Electron

Neutron

Proton

Proton

Nucleus

Nucleus

Helium atom
(mass number = 4)

Carbon atom
(mass number = 12)

Description

The periodic table (see p. 37) is organized to show relationships between different elements as you move up and down a column or across a row. What kinds of trends, you ask? Well, it shows valence electrons, atomic size, electronegativity, and ionization energy, to name just a few. Let's examine the first two trends mentioned: valence electrons and atomic size.

Valence Electrons

Valence electrons = the number of electrons in the outermost shell of an atom

Because the number of electrons in the outermost shell of an atom is directly involved in chemical bonding and determines reactivity, it is handy to know how many are present. Elements in the same column have the same number of valence electrons. Look at the illustration on the facing page. The columns called groups (families) are highlighted in color. Each column is numbered at the top from 1A to 8A. This number tells the number of valence electrons for each element. For example, all the elements in 1A, from hydrogen (H) at the top to francium (Fr) on the bottom, have 1 valence electron. All the elements in 2A have 2 valence electrons, and so on. An illustration of a hydrogen (H) atom is shown. Because hydrogen (H) is in family 1A, it has one valence electron.

Atomic Size

Atoms of different elements are not all the same size. Atomic size often is expressed as the radius of the atom, meaning the distance from the center of the atom to the outermost electron shell. As the atomic radius increases, so does the atomic size, and vice versa. The illustration on the facing page shows two trends: (1) atomic size increases as you move down a group, and (2) atomic size decreases as you move across a row (period) from left to right. Notice that this is a trend and not a hard-and-fast rule. To illustrate this point let's look at an exception to the rule—namely, polonium (Po). Based on the aforementioned rule, you would predict that Po would be smaller than its neighbor to the left, namely bismuth (Bi), but it is not. Once again, we are observing *trends*, so there will always be a few exceptions to the rule.

Practice Problems

Using the illustration on the facing page, determine the number of valence electrons for each of the following elements:

1 Calcium (Ca)

2 Phosphorus (P)

3 Xenon (Xe)

Using the illustration on the facing page, answer the following regarding atomic size:

4 Predict which element is larger, sodium (Na) or chlorine (Cl). Explain.

5 Predict which element is larger, lithium (Li) or potassium (K). Explain.

Answers

1 2   2 5   3 8   4 Sodium (Na) is larger because it is in the same row as chlorine but farther to the left. The rule is that atomic size decreases as you move across a row left to right.   5 Potassium (K) is larger because it is below lithium in the same column, and the rule is that atomic size increases as you move down a column.

Valence electrons

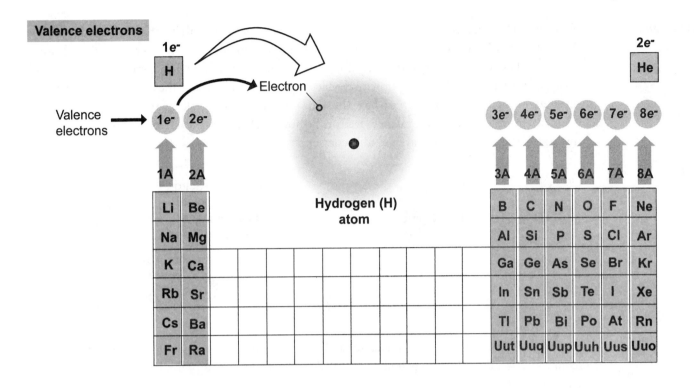

Atomic size

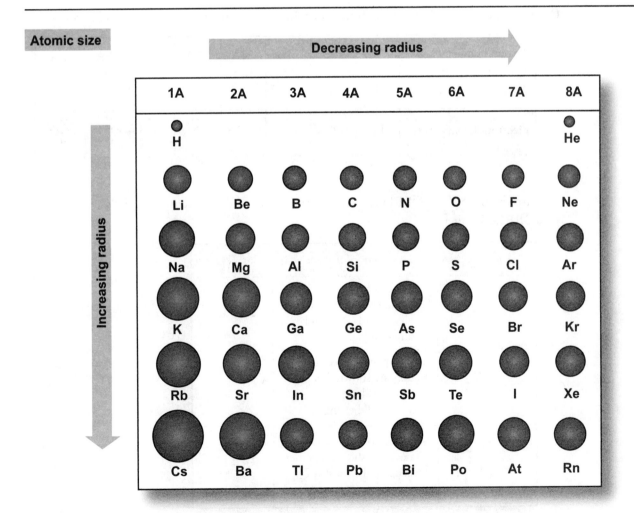

Description

Electron–dot structures, also called Lewis dot structures, refer to a visual shorthand method to show valence electrons. First, the symbol for the element is written, then valence electrons are positioned around the symbol as dots. To be organized, the dots are positioned like points on a compass at north, south, east, and west. Look at the illustration on the facing page. It reminds us of the periodic trend (see p. 44) in which elements in the same group have the same number of valence electrons. For example, all of the elements in group 1A have one valence electron. So how would we write an electron–dot structure for the hydrogen atom shown? Here are the four correct possibilities:

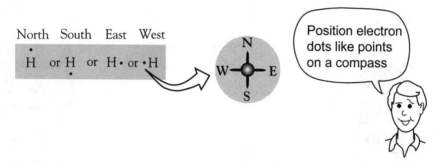

For atoms that have 1–4 valence electrons, dots are positioned as single dots at each point on the compass. After that, dots are paired as double dots. To illustrate this point, let's use an example from groups 1A–8A:

(*Note:* One exception in group 8A is helium, which has 2 valence electrons instead of 8.)

Group number	1A	2A	3A	4A	5A	6A	7A	8A
Valence electrons	1	2	3	4	5	6	7	8
Electron-dot structures	Li·	Be·	·B·	·C·	·N·	·O:	·F:	:Ne:

We already know the CORRECT way to write electron–dot structures, so let's learn from some **INCORRECT** structures:

Incorrect	Description of error	Correct
B:	A pair of electron–dots is shown, but valence electrons 1–4 should all be single.	·B·
Si:	Two pairs of electron–dots are shown, but valence electrons 1–4 should all be single.	·Si·
S:	3 electron–dots should never be shown together because the most is 2 at any one position.	·S:

Valence electrons

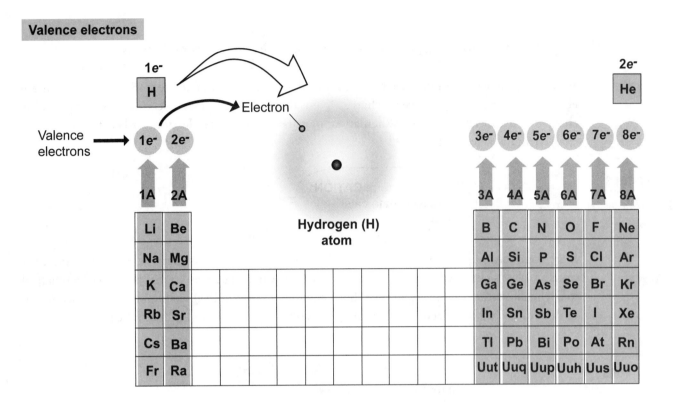

Practice Problems

Using the illustration above, write electron–dot structures for each of the following:

1 Potassium (K)

2 Magnesium (Mg)

3 Aluminum (Al)

4 Phosphorus (P)

5 Bromine (Br)

6 Argon (Ar)

Answers

1 K· 2 Mg· 3 ·Al· 4 ·P· 5 :Br· 6 :Ar:

Ions

> **Ion** = atom that has either *gained* or *lost* one or more electrons

When an atom gains or loses electrons, it is called an **ion** instead of an atom. Recall that atoms are electrically neutral, so they have no charge. Atoms that lose one or more electrons become positively charged and are called **cations**. When an atom gains one or more electrons, it becomes negatively charged and is called an **anion**.

Positive (+) ions (Cations)

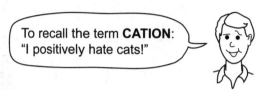

To recall the term **CATION**: "I positively hate cats!"

If an atom loses one electron, its charge is $+1$, if it loses 2 electrons, its charge is $+2$, and so on. Electrons are always negatively charged, so the *loss* of this negative charge results in a positively ($+$) charged ion. The illustration on the facing page shows an atom of sodium (Na). Sodium has only 1 electron in its outermost shell, so it tends to give it up quite easily to become an ion with a $+1$ charge.

Negative (−) ions (Anions)

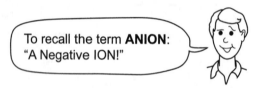

To recall the term **ANION**: "A Negative ION!"

If an atom gains one electron, its charge is -1, if it gains 2 electrons, its charge is -2, and so on. Electrons are always negatively charged, so the *gain* of a negative charge results in a negatively ($-$) charged ion. The illustration on the facing page shows an atom of chlorine. It has 7 electrons in its outermost shell. A full shell contains 8 electrons, so chlorine tends to readily accept an electron to become full. The result is an ion with a -1 charge.

Practice Problems

Answer the following:

1   If an atom *gains* 3 electrons, what is the charge?

2   If an atom *loses* 2 electrons, what is the charge?

3   What is the charge on an atom?

4   What is the difference between a **cation** and an **anion**?

Answers

1 −3   2 +2   3 no charge   4 cations are positively (+) charged and anions are negatively (−) charged.

Positive (+) ions
(cations)

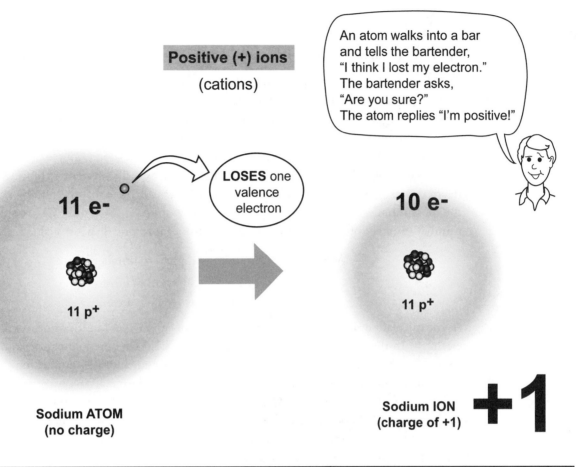

An atom walks into a bar
and tells the bartender,
"I think I lost my electron."
The bartender asks,
"Are you sure?"
The atom replies "I'm positive!"

LOSES one valence electron

11 e⁻

11 p⁺

**Sodium ATOM
(no charge)**

10 e⁻

11 p⁺

**Sodium ION
(charge of +1)** **+1**

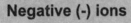

Negative (-) ions
(anions)

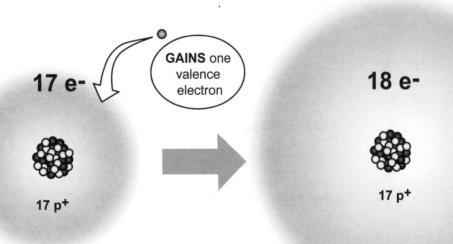

GAINS one valence electron

17 e⁻

17 p⁺

**Chlorine ATOM
(no charge)**

18 e⁻

17 p⁺

**Chloride ION
(charge of -1)** **-1**

Isotopes

> **Isotopes** = atoms of the same element that differ in their number of neutrons

To help us understand isotopes, three different examples are given on the facing page: chlorine (Cl), silicon (Si), and lead (Pb). These elements were selected for two reasons: (1) They represent the three different categories of elements—nonmetals, metalloids, and metals, respectively, and (2) each has a different number of isotopes.

Let's examine each element in turn. Chlorine bleach is a good example of a chlorine-containing substance because it is commonly used in the home as a disinfectant. Chlorine has two isotopes referred to as "chlorine 35" (^{35}Cl) and "chlorine 37" (^{37}Cl). Note that the numbers 35 and 37 refer to the **mass number**. To determine the number of neutrons in ^{35}Cl and ^{37}Cl, we subtract the **atomic number** (number of protons) from the **mass number** (protons + neutrons):

> **Mass number** − atomic number = number of neutrons

$$\text{For } ^{35}Cl: 35 - 17 = 18 \text{ neutrons}$$
$$\text{For } ^{37}Cl: 37 - 17 = 20 \text{ neutrons}$$

As the calculation shows, ^{37}Cl has two more neutrons than ^{35}Cl but the same number of protons and electrons. In the natural environment, the isotope ^{35}Cl occurs predominantly and makes up 75.8% while ^{37}Cl comprises only 24.2%. This helps us understand chlorine's **atomic mass** of 35.45. Because ^{35}Cl constitutes three-fourths of all the isotopic forms of chlorine, it's no surprise that the atomic mass is much closer to 35 than it is to 37. Next, let's examine silicon (Si), which has three isotopic forms. At slightly over 92% in occurrence, the isotope ^{28}Si is by far the most common. Nevertheless, two other forms exist in much smaller proportions, namely ^{29}Si and ^{30}Si. Silicon has many uses in industry, and a silicon chip symbolizes our technological age. From computers to cell phones, a highly pure form of silicon is needed to make these devices. The last example is lead (Pb). The phrase from old cowboy movies, "I'm gonna fill you full of lead!" comes from the fact that bullets often are made from this element. There are four isotopes of lead with ^{208}Pb making up the largest fraction at 52.4%. The percentages of ^{206}Pb and ^{207}Pb are nearly equal, with the smallest fraction represented by ^{204}Pb at 1.4%.

Practice Problems

Using the illustration on the facing page and the formula given above, calculate the number of **neutrons** for each isotope of silicon (Si):

1. ^{28}Si

2. ^{29}Si

3. ^{30}Si

Using the illustration on the facing page, answer the following about the isotopes of lead (Pb):

4. Which isotope has more neutrons: ^{204}Pb or ^{208}Pb? How do you know?

5. How many more neutrons does ^{206}Pb have compared to ^{204}Pb?

Answers

5 ^{206}Pb has two more neutrons than ^{204}Pb; this is determined by subtracting their mass numbers.
4 ^{208}Pb has more neutrons because of its higher mass number
1 $28-14 = 14$ neutrons 2 $29-14 = 15$ neutrons 3 $30-14 = 16$ neutrons

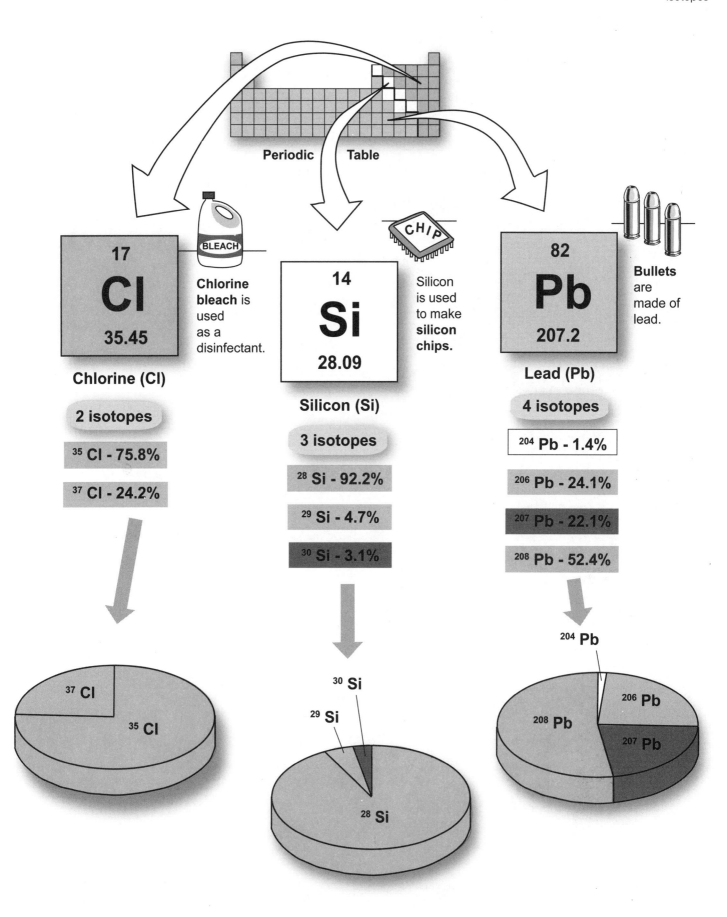

Periodic Table

Chlorine bleach is used as a disinfectant.

17
Cl
35.45

Silicon is used to make **silicon chips.**

14
Si
28.09

Bullets are made of lead.

82
Pb
207.2

Chlorine (Cl)

2 isotopes

^{35} Cl - 75.8%

^{37} Cl - 24.2%

Silicon (Si)

3 isotopes

^{28} Si - 92.2%

^{29} Si - 4.7%

^{30} Si - 3.1%

Lead (Pb)

4 isotopes

^{204} Pb - 1.4%

^{206} Pb - 24.1%

^{207} Pb - 22.1%

^{208} Pb - 52.4%

^{37} Cl

^{35} Cl

^{30} Si

^{29} Si

^{28} Si

^{204} Pb

^{206} Pb

^{208} Pb

^{207} Pb

51

Chemical
Bonds

Octet Rule and Exceptions

Matter always wants to be in its most stable state. But what determines this stability? One measure is an atom's total number of valence electrons. The **octet rule** describes the *general* rule for stability for an atom: 8 electrons in its outermost shell (or 8 valence electrons). Why 8? This is the typical maximum number of electrons that makes most electron shells "full." To remember the octet rule, think of the eight ball used in playing pool.

EIGHT BALL = octet rule

:Ne:

For example, neon (Ne) has 8 valence electrons, so it is very stable.

Are there exceptions to this rule? Yes. For example, hydrogen is the simplest element on the periodic table and has one valence electron. Because it has only one electron shell located close to the nucleus, that shell is considered full when it contains 2 electrons. As a result, hydrogen's "magic number" for stability is 2, not 8. The same is true for helium, which contains 2 electrons in its only electron shell.

Analogy

This octet rule is a driving force for chemical bonding. Unstable atoms combine with other unstable atoms to form a more stable product. As an analogy, let's classify a lonely, single guy who is looking for a spouse as "unstable." Ideally speaking, if he bonds with a suitable companion, and they work hard at their relationship, they have formed a more "stable" couple. Obviously, there are LOTS of exceptions to this rule, but you get the point, right?

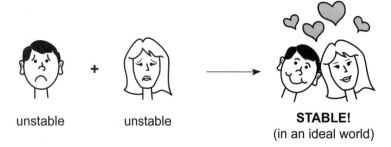

unstable   +   unstable   ⟶   **STABLE!**
(in an ideal world)

Let's apply this analogy to the concept of chemical bonding. Atoms bond with other atoms to form more stable products in several different ways. If they share a pair of electrons, they form a **covalent bond**, but if one atom donates an electron and the other receives it, an **ionic bond** is formed. Both of these types of bond will be discussed as separate topics.

Example of Octet Rule

Let's use the example of an ionic bond in the formation of table salt to understand this further.

Word description

A sodium atom (Na) plus a chlorine atom (Cl) form sodium chloride salt (NaCl)

Chemical equation

Na   **+**   **Cl**   ⟶   **NaCl**

Visual equation

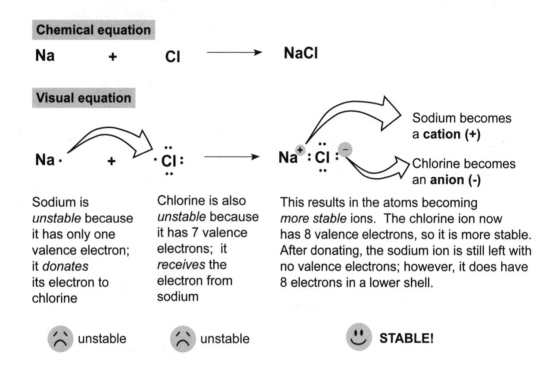

Sodium becomes a **cation (+)**

Chlorine becomes an **anion (-)**

Sodium is *unstable* because it has only one valence electron; it *donates* its electron to chlorine

Chlorine is also *unstable* because it has 7 valence electrons; it *receives* the electron from sodium

This results in the atoms becoming *more stable* ions. The chlorine ion now has 8 valence electrons, so it is more stable. After donating, the sodium ion is still left with no valence electrons; however, it does have 8 electrons in a lower shell.

unstable   unstable   STABLE!

Practice Problems

On a separate sheet of paper, answer the following:

1.   Alkali metals (1A) such as lithium and sodium each have 1 valence electron. Based on what you know about chemical bonding, predict if they are more chemically reactive or less chemically reactive. Explain.

2.   The noble gases (8A) such as neon and argon all have 8 valence electrons. Based on what you know about chemical bonding, predict if they are more reactive or less reactive. Explain.

3.   Calcium has 2 valence electrons. If it formed an ionic bond, would it more likely be an electron donor or a recipient? Explain.

4.   Oxygen has 6 valence electrons. How many more does it need to have a full outermost shell?

Answers

1 Alkali metals need 8 valence electrons to be stable, but have only 1, so they are unstable. This makes them more chemically reactive because they have to lose an electron to become more stable. 2 Noble gases have a full outermost shell with 8 valence electrons, so they are very stable and less chemically reactive. They do not have to bond to become stable. 3 Calcium has only 2 valence electrons but needs 8, so it would be easier for it to donate 2 than receive 6. Thus, it would be an electron donor. 4 Oxygen needs 8 valence electrons to have a full outermost shell. It already has 6, so it would need 2 more. Thus, it would be an electron acceptor.

Description

Atoms bond to be in a more stable state. We're going to examine two common types of bonds: **ionic bonds** and **covalent bonds**.

Ionic bonds = one atom *donates* its electron; the other *receives* that same electron	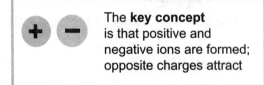 The **key concept** is that positive and negative ions are formed; opposite charges attract

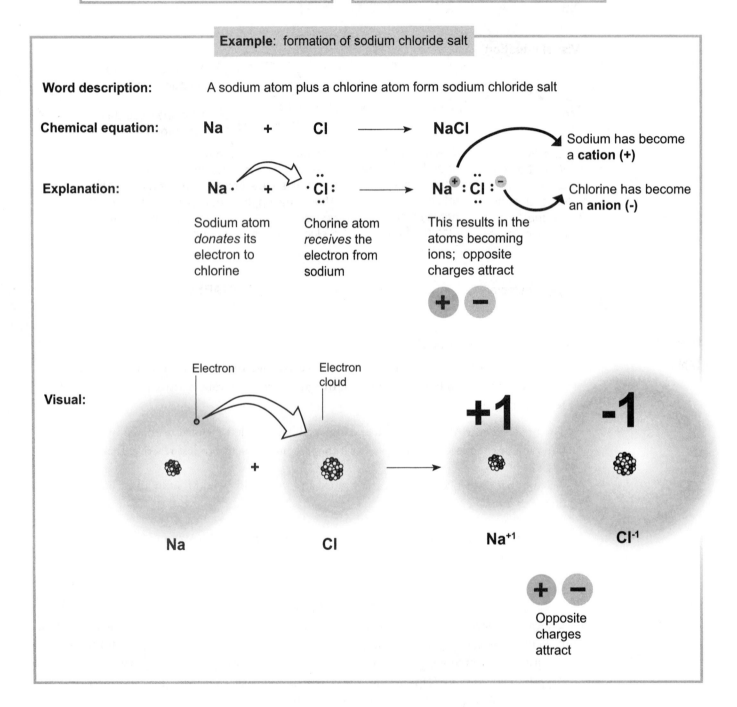

Example: formation of sodium chloride salt

Word description: A sodium atom plus a chlorine atom form sodium chloride salt

Chemical equation: Na   +   Cl   ⟶   NaCl

Sodium has become a **cation (+)**

Chlorine has become an **anion (-)**

Explanation: Na·   +   ·Cl: ⟶   Na⁺ : Cl : ⁻

Sodium atom *donates* its electron to chlorine

Chorine atom *receives* the electron from sodium

This results in the atoms becoming ions; opposite charges attract

Visual:

Electron

Electron cloud

Na   Cl   Na⁺¹   Cl⁻¹

+1   -1

Opposite charges attract

Covalent bonds	= **shared pair** of electrons (like a couple in love, the word "love" is scrambled in the word "covalent")

 The **key concept** is that a covalent bond links two atoms together like two people locking arms

Example: formation of hydrogen gas

Word description: A hydrogen atom plus another hydrogen atom forms hydrogen gas

Chemical equation:

$$H \quad + \quad H \quad \longrightarrow \quad H_2$$

Shared pair

Line is a symbol for a covalent bond

Explanation:

$$H\cdot \quad + \quad \cdot H \quad \longrightarrow \quad H\!:\!H \quad = \quad H_2 \quad = \quad H\!-\!H$$

Hydrogen atom *shares* its valence electron with ...

... another hydrogen atom

This results in a shared pair of electrons or a covalent bond

Electrons in overlapping electron cloud indicate a covalent bond

Visual:

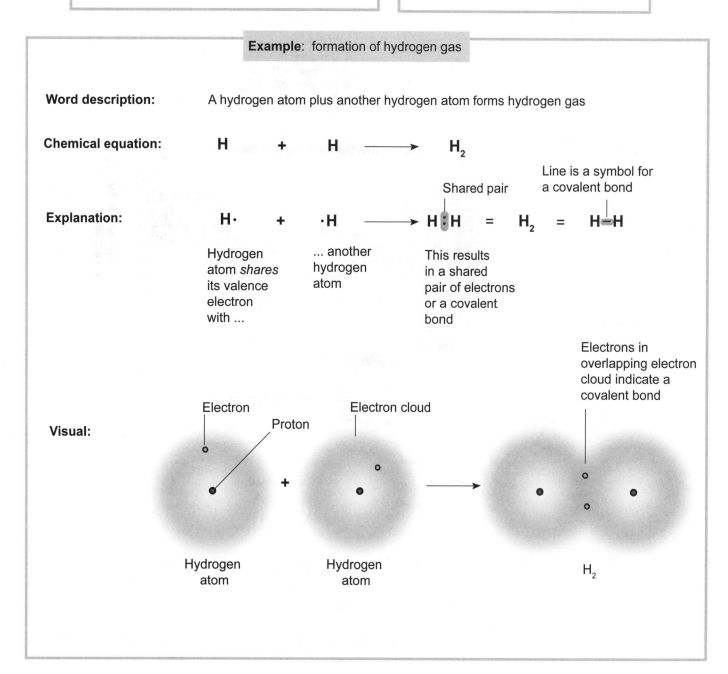

Electron

Proton

Electron cloud

Hydrogen atom

Hydrogen atom

H_2

Electronegativity

Electronegativity is the property of an atom to pull a shared pair of electrons toward itself. It is measured on the Pauling scale (0–4), which evaluates every element relative to fluorine, as it has the highest electronegativity at 4.0. In contrast, the metal cesium (Cs) has the lowest electronegativity at 0.7. The trend on the periodic table is that electronegativity increases as you move up a column and across a row from left to right. As a general rule, nonmetals have a higher degree of electronegativity than metals. Why? It's partly because of their smaller size, which leads to a stronger "pull" between the negatively charged valence electrons and the positively charged nucleus. Here are the five elements with the highest electronegativities and their values:

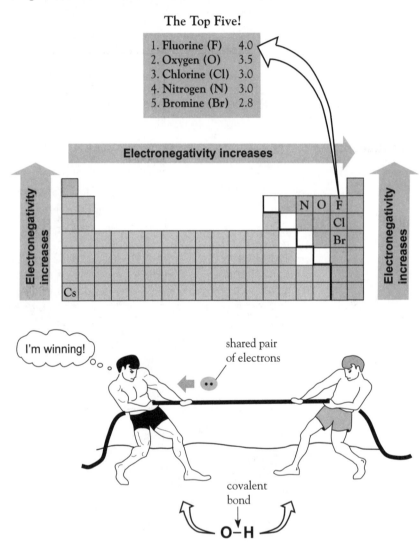

The Top Five!

1. Fluorine (F)	4.0
2. Oxygen (O)	3.5
3. Chlorine (Cl)	3.0
4. Nitrogen (N)	3.0
5. Bromine (Br)	2.8

Electronegativity increases

Electronegativity increases

Electronegativity increases

I'm winning!

shared pair of electrons

covalent bond

O–H

As an analogy, any time two different atoms are covalently bonded together, it's like a **tug of war**. Recall that a covalent bond is a shared pair of electrons between two atoms. The atom with the greater electronegativity will pull harder on the shared pair of electrons. Consider in the example shown above (O–H), where oxygen (O) is covalently bonded to hydrogen (H). Oxygen (3.5) has the greater electronegativity and is like the stronger man in black shorts. Hydrogen (2.1) has the weaker electronegativity and is like the weaker man in colored shorts. In this case, oxygen wins the tug of war. Now let's apply the concept of electronegativity to bond polarity.

Bond Polarity

The two types of covalent bonds are: (1) **nonpolar covalent**, and (2) **polar covalent**. The term *nonpolar* means no electrical charge, and the term *polar* means that the bond has an electrical charge.

Nonpolar covalent bonds have the following features:

- Often occur between the same atoms

- Electronegativity difference between bonded atoms is < 0.5

- Electrons are shared equally between the atoms

As an example, consider one molecule of chlorine gas (Cl_2) or **Cl-Cl**

Predictably, the difference in electronegativity is zero (3.0–3.0) because it is the same atom bonded to itself. This means that the electrons are shared equally because one atom does not pull on them more than the other. The result is that the chlorine-chlorine covalent bond is nonpolar.

Polar covalent bonds have the following features:

- Always occur between different atoms

- Electronegativity difference between bonded atoms is 0.5 to 1.9

- Electrons are NOT shared equally between the atoms

As an example, consider one molecule of hydrochloric acid (HCl) or **H-Cl**

The difference in electronegativity is 0.9 (3.0–2.1). This means that the chlorine atom (3.0) pulls more on the shared pair of electrons than does the hydrogen (2.1). The shared pair of electrons is not shared equally, so it results in a molecule with a partial charge. Chlorine will become partially negative, and hydrogen will become partially positive. In short, chlorine wins the tug of war.

When a molecule has a separation of charge, as with any polar covalent bond, it is called a **dipole** (*two poles*). Think of a dipole as a bar magnet that has a **positive** (+) **end** and a **negative** (−) **end**.

A **dipole** is like a bar magnet

Practice Problems

Classify each bond as either *nonpolar covalent* or *polar covalent*. Electronegativity values are given in parentheses.

1 C–S	2 H–H	3 N–H
(2.5) (2.5)	(2.1) (2.1)	(3.0) (2.1)

Answers

1 The difference in electronegativity is 0 (2.5–2.5), so it is a nonpolar covalent bond.
2 The difference in electronegativity is 0 (2.1–2.1) so it is a nonpolar covalent bond.
3 The difference in electronegativity is 0.9 (3.0–2.1), so it is a polar covalent bond.

We will examine the following three intermolecular forces: **dipole–dipole interactions**, **hydrogen bonds**, and **dispersion forces**. These forces occur within liquids and solids but not gases because their molecules are spread too far apart.

Dipole–dipole Interactions

Key Points

- They are found between **polar molecules.**
- Atoms in polar molecules are two different nonmetals.
- The partially positive end of one molecule is attracted to the partially negative end of another molecule.
- They are a relatively strong force.

The example on the facing page shows a molecule of hydrochloric acid (HCl) is attracted to another molecule of HCl. It fits all the key points listed above because HCl is a polar molecule composed of two nonmetals with a partially positive charge ($\delta +$) on the hydrogen end and a partially negative charge ($\delta -$) on the chlorine end. This partial charge means the chlorine end on one molecule is always attracted to the hydrogen end of another molecule. Dipole-dipole interactions are relatively strong forces.

Hydrogen Bonds

Key Points

- A dipole–dipole interaction occuring between a **H atom** bonded to an **O, N, or F atom**.
- Partially positive end of H in one molecule is attracted to partially negative O, N, or F of another molecule.
- They are the **strongest** of the three intermolecular forces.

The example on the facing page shows a molecule of water (H_2O) attracted to another molecule of H_2O. It fits all the key points listed above because H_2O is a polar molecule composed of hydrogen and oxygen atoms. The hydrogen ends of the molecule have a partially positive charge ($\delta +$), and the oxygen end has a partially negative charge ($\delta -$). Note that oxygen (O), and nitrogen (N) are both highly electronegative, resulting in polar bonds. This polarity difference means that the hydrogen end on one molecule is always attracted to the oxygen end of another molecule. Hydrogen bonds are the strongest of the three intermolecular forces. In biomolecules, they help maintain the three-dimensional shape of proteins, which is important to their function, and they hold the "bases" together in DNA.

Dispersion Forces

Key Points

- They are found between **nonpolar molecules.**
- Atoms in nonpolar molecules are the same element or have similar electronegativity.
- The partial positive end of one molecule is attracted to the partial negative end of another molecule.
- They are the **weakest** of the three intermolecular forces.

The example on the facing page shows a molecule of chlorine gas (Cl_2) being attracted to another molecule of Cl_2. It fits all the key points listed above because Cl_2 is a nonpolar molecule consisting of the same atom bonded to itself. Typically, this results in a molecule that has no charge. But random shifts in the distribution of the electrons results in a *temporary* polar molecule (though it lasts for only a fraction of a second). This short-term charge means that the negative end of one chlorine gas molecule can be attracted to the positive end of another chlorine gas molecule. Note that this is the only type of attractive force between nonpolar molecules. As you might guess, dispersion forces are the weakest of the three forces presented here.

Dipole-dipole interactions

Example: attraction between molecules of hydrochloric acid (HCl)

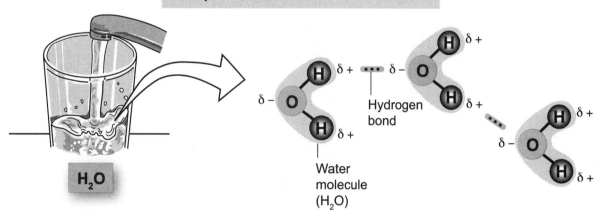

$\delta +$ H—Cl $\delta -$ ••• $\delta +$ H—Cl $\delta -$ ••• $\delta +$ H—Cl $\delta -$

Dipole-dipole interaction

HYDROCHLORIC ACID

HCl

Molecule of hydrochloric acid (HCl)

Hydrogen bonds

Example: attraction between water molecules

H₂O

$\delta -$ O, H $\delta +$, H $\delta +$ ••• $\delta -$ O, H $\delta +$, H $\delta +$ ••• $\delta -$ O, H $\delta +$, H $\delta +$

Hydrogen bond

Water molecule (H₂O)

Dispersion forces

Example: attraction between molecules of chlorine gas

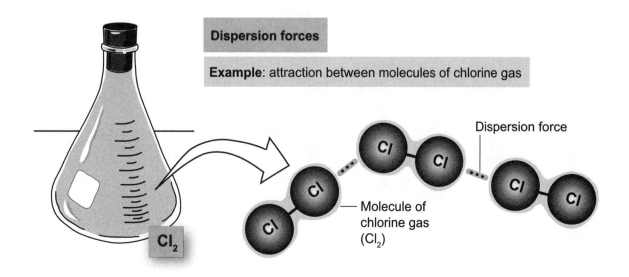

Dispersion force

Cl₂

Molecule of chlorine gas (Cl₂)

Chemical Quantities

Description

The **mole (mol)** is a convenient *unit* of measurement used by chemists to deal with a large number of small substances such as atoms, molecules, and compounds. We deal with units of measurement every day that relate to relatively small quantities such as a dozen. No matter what we purchase, if we buy a dozen, we know we are getting 12. Similarly, if we buy a case (24) of soda pop or a gross (144) of pencils, the amount is always a fixed number. Because chemists deal with atoms and other extremely small items, they need a much larger fixed number than 12, 24, or 144. A mole is based on **Avogadro's number** and equals 602 billion trillion items or 6.02×10^{23}.

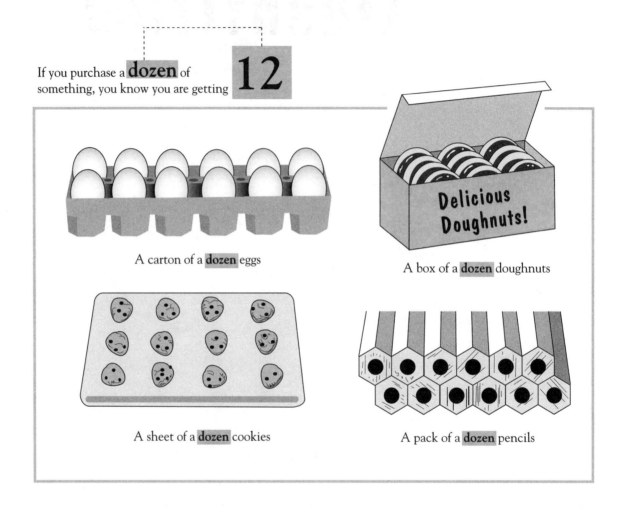

If you purchase a **dozen** of something, you know you are getting **12**

A carton of a **dozen** eggs

A box of a **dozen** doughnuts

Delicious Doughnuts!

A sheet of a **dozen** cookies

A pack of a **dozen** pencils

Common Units Like the Mole

UNIT	AMOUNT
Dozen	12
Case	24
Gross	144
Mole	6.02×10^{23}

The mole is about **"HOW MANY,"** not about "how heavy"!

A mole can't be measured directly with any device, like temperature is with a thermometer. Instead, we weigh the amount of a substance equal to a mole. The mass of a mole of any substance is equal to its **molar mass**.

Remember, there is no **mole**–*ometer*!

Let's look at a **mole** of four different substances: a *liquid*, a *metal*, a *gas*, and an *organic compound*.

6.02 × 10²³ items

(602,000,000,000,000,000,000,000)

602 billion trillion items!

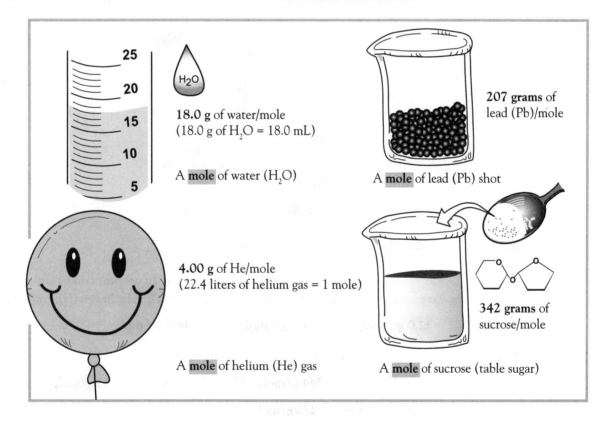

18.0 g of water/mole
(18.0 g of H₂O = 18.0 mL)

A **mole** of water (H₂O)

207 grams of lead (Pb)/mole

A **mole** of lead (Pb) shot

4.00 g of He/mole
(22.4 liters of helium gas = 1 mole)

A **mole** of helium (He) gas

342 grams of sucrose/mole

A **mole** of sucrose (table sugar)

Description

Molar mass is defined as the mass of one mole of any substance and is expressed in units of grams/mole.

$$\text{Molar mass} = \text{grams (g)} / \text{mole (mol)}$$

For example, in the previous module it was stated that one mole of the element lead (Pb) weighed 207 grams.

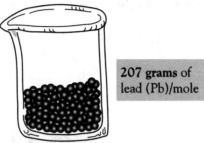

207 grams of lead (Pb)/mole

A **mole** of lead (Pb) shot

How was this determined? The mass of a mole is equal to a substance's atomic mass in grams. From the periodic table, we can identify lead's atomic mass, 207. Therefore, lead's molar mass equals 207 grams/mole.

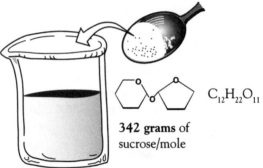

$C_{12}H_{22}O_{11}$

342 grams of sucrose/mole

A **mole** of sucrose (table sugar)

How do you calculate the molar mass of a compound like sucrose?

Step 1   Obtain the molecular formula for sucrose—$C_{12}H_{22}O_{11}$—then use the periodic table to reference molar masses of its component elements, namely carbon (C), hydrogen (H), and oxygen (O):

$$C = \boxed{12.0 \text{ g} / \text{mol,}} \quad H = \boxed{1.0 \text{ g} / \text{mol,}} \quad O = \boxed{16.0 \text{ g} / \text{mol}}$$

Step 2   Calculate the grams for total moles of each element:

$$C = 12 \times 12 \text{ g} / \text{mol} = \boxed{144 \text{ g/mol,}} \quad H = 22 \times 1 \text{ g/mole} = \boxed{22 \text{ g/mol,}}$$

$$O = 11 \times 16 \text{ g} / \text{mole} = \boxed{176 \text{ g/mol}}$$

Step 3   Add the molar masses calcuated in step 2 to determine the compound's molar mass:

$$144 \text{ g/mol} + 22 \text{ g/mol} + 176 \text{ g/mol} = \boxed{342 \text{ grams of sucrose/mol}}$$

Practice Problems

Calculate the molar mass for each of the following substances:

1 NaCl

2 $Ca_3(PO_4)_2$

3 $C_6H_{12}O_6$

4 $MgSO_4$

5 CH_3CH_2OH

Answers

1 58.44 g/mole 2 310.18 g/mole 3 180.16 g/mole 4 120.37 g/mole 5 46.07 g/mole

Chemical
Equations

Physical Change

A **physical change** does *not* involve a chemical reaction. Instead, a physical property of a substance is altered, but not its chemical formula. These changes often involve a change of state. The example on the facing page shows the freezing of water—a change of state from a liquid to a solid. The chemical formula for water is H_2O. Notice that this formula, H_2O, remains as H_2O despite the change in state. Other physical changes may alter the shape, size, or texture of a substance. Here are some other examples:

- **Folding a sheet of paper in half** (size change)
- **Crushing an aluminum can** (shape and size change)
- **Sanding a rough piece of wood** (texture change)
- **Boiling water** (change of state from a liquid state to a gaseous state)

Chemical Change

A **chemical change** involves a chemical reaction in which one or more new substances are produced. Evidence of a chemical reaction occurring includes the production of energy (heat, light, sound), a gas (bubbles), a solid (precipitate), or a color change. The example on the facing page shows nails rusting. The iron (Fe) in the nails undergoes a chemical reaction in the presence of oxygen in the air to produce a new substance, rust, or iron oxide (Fe_2O_3). Other examples of chemical changes include:

- **Burning paper** (energy is released in the form of heat and light)
- **Mixing an antacid tablet with an acid** (carbon dioxide gas bubbles form)
- **Brewing beer** (carbon dioxide gas bubbles form)
- **Changing color of apple slices** (color change from white to brown)

Practice Problems

Classify the following as either a physical change or a chemical change:

1 Breaking a matchstick in half

2 Burning a matchstick

3 Breaking glass

4 Melting glass

5 Heating sugar until it caramelizes

Answers

1 Physical change 2 Chemical change 3 Physical change 4 Physical change 5 Chemical change

Physical change
Example: **freezing of water**

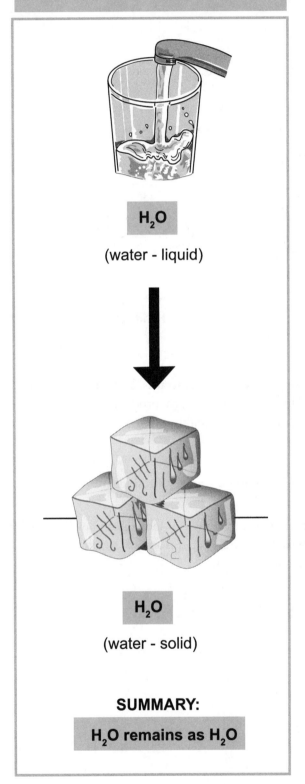

H₂O

(water - liquid)

H₂O

(water - solid)

SUMMARY:

H₂O remains as H₂O

Chemical change
Example: **nails rusting**

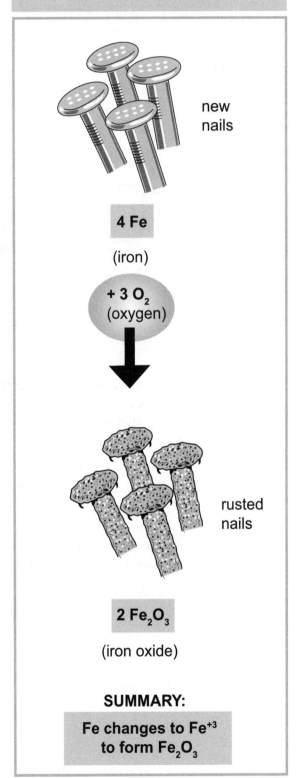

new nails

4 Fe

(iron)

+ 3 O₂
(oxygen)

rusted nails

2 Fe₂O₃

(iron oxide)

SUMMARY:

**Fe changes to Fe⁺³
to form Fe₂O₃**

Description　Knowing the terms and symbols used in chemical equations is essential to understanding chemistry. Let's use making pancakes as an analogy. Mixing ingredients together in a recipe is similar to what occurs in some chemical reactions. By combining two substances, then heating the batter, you form something new.*

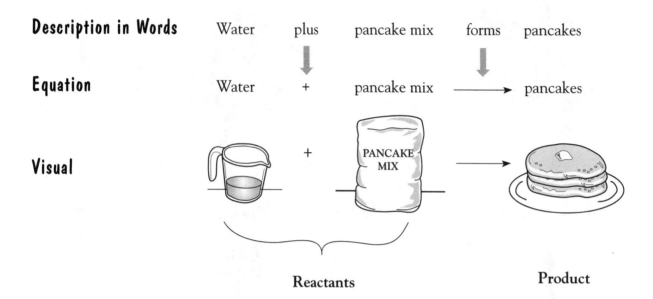

Description in Words　Water　plus　pancake mix　forms　pancakes

Equation　Water　+　pancake mix ⟶ pancakes

Visual

Reactants

Product

In the above example, the *water* and the *pancake mix* are the **reactants**. They always are written to the left of the arrow. After the batter is heated in a frying pan, the final **product**, pancakes, is formed. The product always is written to the right of the arrow. Knowing this, let's write the general formula for any chemical equation:

Reactant #1　+　Reactant #2　⟶　Product(s)

* This analogy works only in the broadest sense of combining two substances to make something new. When water is combined with pancake mix, the resulting batter is actually a mixture. When the batter is heated, it undergoes a chemical change to form pancakes.

Practice Problems

Identify the **REACTANTS** in the following chemical equations:

1 $H_2CO_3 \longrightarrow H_2O + CO_2$

2 $BaCl_2 + Na_2SO_4 \longrightarrow 2NaCl + BaSO_4$

3 $C + O_2 \longrightarrow CO_2$

Identify the **PRODUCTS** in the following chemical equations:

4 $CaCO_3 + 2HCl \longrightarrow CaCl_2 + CO_2 + H_2O$

5 $2KClO_3 \longrightarrow 2KCl + 3O_2$

6 $CH_4 + 2O_2 \longrightarrow CO_2 + 2H_2O$

Change the following word descriptions into chemical equations:

7 Hydrochloric acid plus sodium hydroxide forms sodium chloride plus water.

8 Glucose plus oxygen forms carbon dioxide plus water.

9 Chlorine plus sodium bromide forms sodium chloride plus bromine.

Answers

9 $Cl_2 + 2NaBr \longrightarrow 2NaCl + Br_2$

8 $C_6H_{12}O_6 + 6O_2 \longrightarrow 6CO_2 + 6H_2O$

7 $HCl_{(aq)} + NaOH \longrightarrow NaCl + H_2O$

6 CO_2, H_2O = carbon dioxide, water

5 KCl, O_2 = potassium chloride, oxygen

4 $CaCl_2, CO_2, H_2O$ = calcium chloride, carbon dioxide, water

3 C, O_2 = carbon, oxygen

2 $BaCl_2, Na_2SO_4$ = barium chloride, sodium sulfate

1 H_2CO_3 = carbonic acid

Chemical Reactions

Description

Chemical reactions have predictable patterns depending on their type. Let's examine three basic types of chemical reactions: *combination reactions*, *decomposition reactions*, and *replacement reactions*.

Combination Reactions

Two or more reactants combine to form a **single** product; the opposite of a decomposition reaction.

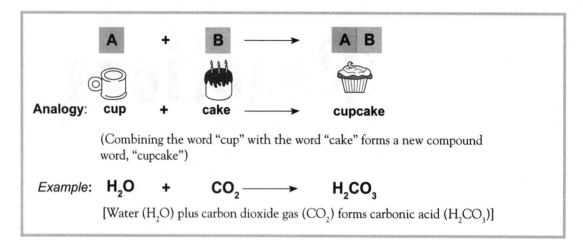

Analogy: cup   +   cake   ⟶   cupcake

(Combining the word "cup" with the word "cake" forms a new compound word, "cupcake")

Example: H_2O + CO_2 ⟶ H_2CO_3

[Water (H_2O) plus carbon dioxide gas (CO_2) forms carbonic acid (H_2CO_3)]

Decomposition Reactions

A compound breaks down into its component parts—the opposite of a combination reaction

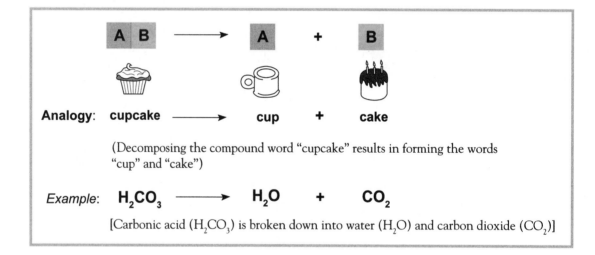

Analogy: cupcake   ⟶   cup   +   cake

(Decomposing the compound word "cupcake" results in forming the words "cup" and "cake")

Example: H_2CO_3 ⟶ H_2O + CO_2

[Carbonic acid (H_2CO_3) is broken down into water (H_2O) and carbon dioxide (CO_2)]

Replacement Reactions

Like changing dance partners, components recombine to form new products. There are two types: **single replacement** and **double replacement**.

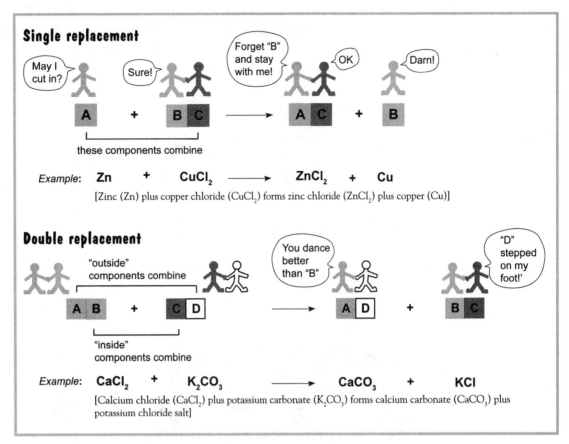

Single replacement

May I cut in?

Sure!

Forget "B" and stay with me!

OK

Darn!

A + B C ⟶ A C + B

these components combine

Example: Zn + CuCl$_2$ ⟶ ZnCl$_2$ + Cu

[Zinc (Zn) plus copper chloride (CuCl$_2$) forms zinc chloride (ZnCl$_2$) plus copper (Cu)]

Double replacement

"outside" components combine

You dance better than "B"

"D" stepped on my foot!'

A B + C D ⟶ A D + B C

"inside" components combine

Example: CaCl$_2$ + K$_2$CO$_3$ ⟶ CaCO$_3$ + KCl

[Calcium chloride (CaCl$_2$) plus potassium carbonate (K$_2$CO$_3$) forms calcium carbonate (CaCO$_3$) plus potassium chloride salt]

Practice Problems

Classify each of the following reactions as *combination*, *decomposition*, *single replacement*, or *double replacement*:

1 $2Mg + O_2 \longrightarrow 2MgO$

2 $NaCl + AgNO_3 \longrightarrow NaNO_3 + AgCl$

3 $Cl_2 + 2NaBr \longrightarrow 2NaCl + Br_2$

4 $SO_2 + H_2O \longrightarrow H_2SO_3$

Answers

1 Combination   2 Double replacement   3 Single replacement   4 Combination

77

Description

Oxidation–reduction reactions (redox reactions) are a very common type of chemical reaction that involve the transfer of electrons between a pair of substances. One substance acts as the electron *donor* and the other acts as the electron *receiver*. The electron donor is referred to as the **reducing agent**, the electron receiver is the **oxidizing agent**. By definition, **oxidation** is loss of electrons and **reduction** is gain of electrons.

Concept

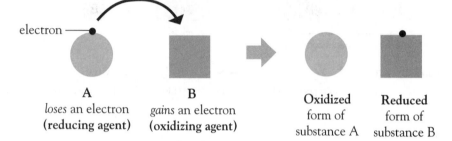

electron →

A
loses an electron
(reducing agent)

B
gains an electron
(oxidizing agent)

Oxidized
form of
substance A

Reduced
form of
substance B

Substance A loses its electron to substance B, so A gets oxidized and B gets reduced.

Analogy

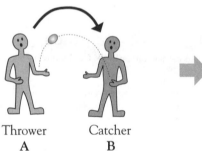

Thrower
A

Catcher
B

Thrower **A**
is now **oxidized**

Catcher **B**
is now **reduced**

Like two people playing catch, redox always occurs in pairs. One is the *donor* and the other is the *recipient*.

Example

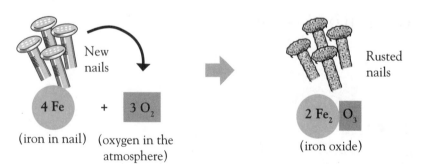

New
nails

$4\ Fe$

$+$

$3\ O_2$

(iron in nail)

(oxygen in the
atmosphere)

Rusted
nails

$2\ Fe_2\ O_3$

(iron oxide)

Rusting is an example of a redox reaction. Electrons from Fe are transferred to O_2. The Fe is oxidized to Fe^{+3} and the O is reduced to form O^{-2}. The resulting compound is Fe_2O_3.

Other Examples

Combustion

Combustion reactions are redox reactions. When gasoline is mixed with oxygen, a combustable mixture is produced. In this reaction, oxygen serves as the oxidizing agent and gasoline is the reducing agent.

Bleaching

Bleaches are oxidizing agents. When cleaning clothes, they remove electrons from colored compounds, resulting in colorless compounds.

Batteries

Batteries generate electricity through redox reactions.

Mnemonic

GER!

"**LEO** says **GER**"

Loss of **E**lectrons is **O**xidation
Gain of **E**lectrons is **R**eduction

Practice Problems

For the following reactions, indicate which substance is being reduced and which is being oxidized. Also identify the reducing agent and the oxidizing agent.

1 Cu^{+2} + Zn $\longrightarrow$ Zn^{+2} + Cu

2 H_2 + F_2 $\longrightarrow$ 2HF

3 $2Mg^{+2}$ + O_2 $\longrightarrow$ 2MgO

Answers

	Being reduced	Being oxidized	Reducing agent	Oxidizing agent
1	Cu^{+2}	Zn	Zn	Cu^{+2}
2	F_2	H_2	H_2	F_2
3	O_2	Mg	Mg	O_2

In almost all chemical reactions, heat is either absorbed or released. When considering this *heat of reaction*, it forces us to examine the two possibilities, **exothermic reactions** and **endothermic reactions**.

EXOthermic reaction **Heat** (The icon to the left illustrates the concept that heat is given off to its immediate surroundings)

"Exothermic" means "outer heat," which correctly describes a reaction in which heat is released to its immediate surroundings. A reaction is considered to be exothermic if it meets the following standards:

- **Releases heat** to the immediate surroundings
- **Increases** the **temperature** of the immediate surroundings
- Occurs **spontaneously**

The example on the facing page shows a hot pack that is used to relax strained muscles and increase circulation. It contains two compartments: one holds water, and the other contains ammonium nitrate (NH_4NO_3), a solid. When squeezed, the compartments break open and the NH_4NO_3 dissolves in water. This releases heat into the solution and increases the temperature of the hot pack. Notice that heat is written on the product side of the equation.

$$NH_4NO_{3(s)} \xrightarrow{H_2O} NH_4NO_{3(aq)} \text{ (and heat)}$$

Other examples of **exothermic reactions** include:
- Burning sugar (as in toasting a marshmallow)
- Freezing water into ice cubes (a freezer removes heat)

ENDOthermic reaction **Heat** (The icon to the left illustrates the concept that heat is absorbed from its immediate surroundings)

"Endothermic" means "inner heat," which aptly describes a reaction in which heat is absorbed from its immediate surroundings. A reaction is considered endothermic if it meets the following standards:

- **Absorbs heat** from the immediate surroundings
- **Decreases temperature** of the immediate surroundings
- Does *not* occur spontaneously

The example on the facing page shows a cold pack that is used to decrease redness, swelling, and pain at the site of an injury. It has the same structure as a hot pack except that it uses calcium chloride ($CaCl_2$) instead of NH_4NO_3. When squeezed, the compartments break open and the $CaCl_2$ dissolves in water. As heat is absorbed, the temperature of the cold pack decreases. Notice that "heat" is written on the *reactant* side of the equation.

$$\text{(heat and) } CaCl_{2(s)} \xrightarrow{H_2O} CaCl_{2(aq)}$$

Other examples of **endothermic reactions** include:
- Producing sugar through photosynthesis (without heat, the process does not occur)
- Melting ice cubes into water (without heat, melting does not occur)

EXOthermic reaction	ENDOthermic reaction
Example: **HOT pack**	*Example*: **COLD pack**

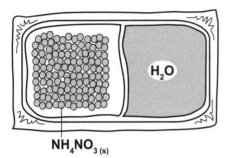

$NH_4NO_{3\,(s)}$

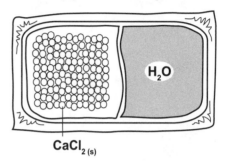

$CaCl_{2\,(s)}$

① Before using the hot pack, the NH_4NO_3 and the water are in separate compartments

① Before using the cold pack, the $CaCl_2$ and the water are in separate compartments

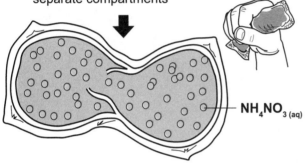

$NH_4NO_{3\,(aq)}$

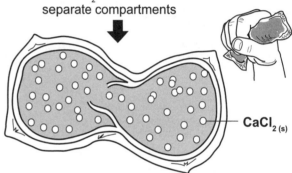

$CaCl_{2\,(s)}$

② Squeezing the pack breaks open the compartments; NH_4NO_3 dissolves in water

② Squeezing the pack breaks open the compartments; $CaCl_2$ dissolves in water

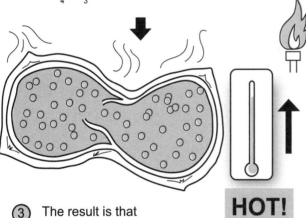

HOT!

COLD!

③ The result is that heat is *PRODUCED*

③ The result is that heat is *REMOVED*

$$NH_4NO_{3\,(s)} \xrightarrow{H_2O} NH_4NO_{3\,(aq)} \text{ (and heat)}$$

$$\text{(heat and) } CaCl_{2\,(s)} \xrightarrow{H_2O} CaCl_{2\,(aq)}$$

Gases

Description

Gases are all around us—in the natural world and in industry. Let's consider some examples. In the natural world, the air we breathe is a mixture of gases such as nitrogen (N_2), oxygen (O_2), carbon dioxide (CO_2), and water vapor ($H_2O_{(g)}$). The cells of living organisms depend on oxygen to sustain life while producing carbon dioxide as a waste product. That's why an airstone in an aquarium bubbles oxygen gas into the water for the fish. In industry, the gasoline engines in cars, trucks, lawn mowers, and other machines produce exhaust gases such as carbon dioxide. Smoke stacks from industrial plants emit gases into the atmosphere. If an old gas furnace in a home is damaged, it can leak carbon monoxide (CO) gas, which can kill people by suffocating them. Helium (He) gas is used to fill balloons. In fact, anything that is inflated—tires, air mattresses, or beach balls—is filled with gas. To be sure, gases are vitally important and are found everywhere.

Properties of Gases

Here is a summary of the properties of a gas:

- **Arrangement:** particles are far apart so they do not directly interact with each other
- **Movement:** particles move quickly, randomly, and in a linear path
- **Shape:** particles take the shape of the container
- **Volume:** particles fill all the space in the container

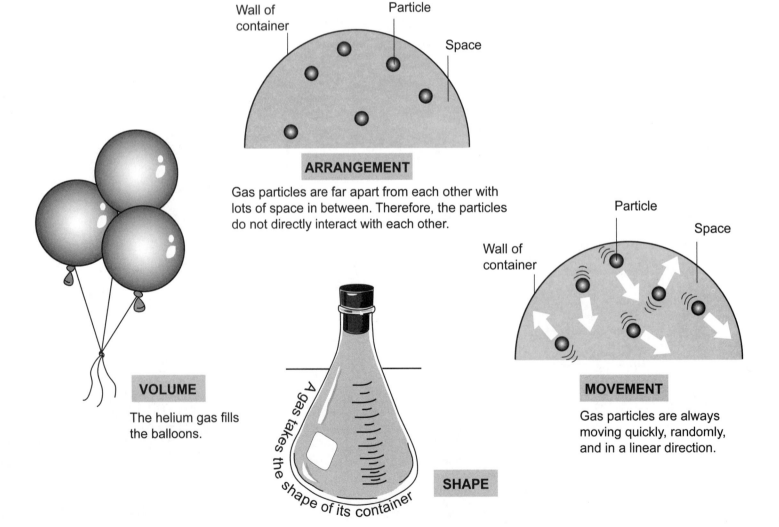

Wall of container   Particle   Space

ARRANGEMENT

Gas particles are far apart from each other with lots of space in between. Therefore, the particles do not directly interact with each other.

VOLUME

The helium gas fills the balloons.

A gas takes the shape of its container

SHAPE

Particle   Space

Wall of container

MOVEMENT

Gas particles are always moving quickly, randomly, and in a linear direction.

Other Properties of Gases

Here are two other properties of a gas:

- **Compressibility:** The large amount of space between particles makes gases easily compressible

- **Movement is proportional to temperature:** An increase in temperature causes increased kinetic energy, leading to increased movement, and vice versa

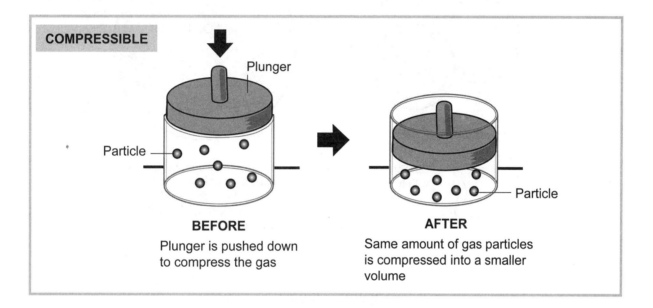

COMPRESSIBLE

Plunger

Particle

Particle

BEFORE
Plunger is pushed down to compress the gas

AFTER
Same amount of gas particles is compressed into a smaller volume

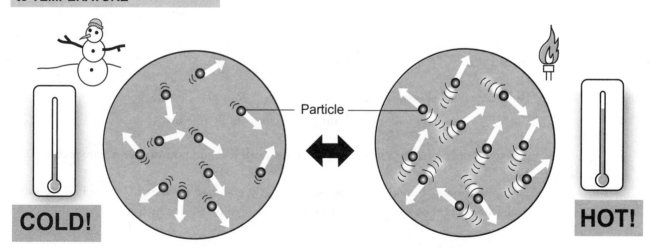

MOVEMENT is PROPORTIONAL to TEMPERATURE

Particle

COLD!

HOT!

DECREASED TEMPERATURE
At *lower* temperatures, gas particles move *slower* because they have less kinetic energy.

INCREASED TEMPERATURE
At *higher* temperatures, gas particles move *faster* and exert a stronger force against their container because they have more kinetic energy.

Description

Gas particles have kinetic energy and are in constant motion; they bounce off the walls of their container and each other. In doing so, they exert a force called gas pressure. It's the force that keeps balloons, tires, and air mattresses firm and inflated. You are already familiar with gas pressure from everyday experiences.

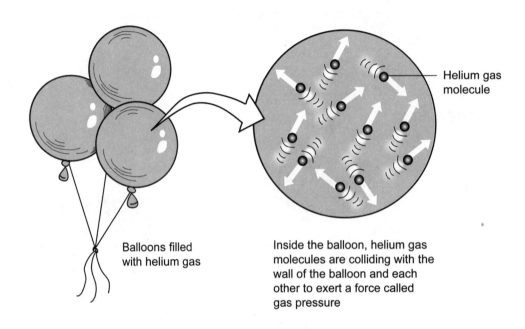

Balloons filled with helium gas

Helium gas molecule

Inside the balloon, helium gas molecules are colliding with the wall of the balloon and each other to exert a force called gas pressure

Units

When measuring gas pressure, the three commonly used units are:

Units	Abbreviation
millimeters of mercury	mm Hg
Torr	Torr
atmosphere	atm

The conversion between the three commonly used units listed in the table:

1 atm = 760 mm Hg = 760 Torr

Less commonly used units in science include **pascals** (Pa) and **pounds per square inch** (psi).

The significance of the number "760" is that it is the standard measure of atmospheric pressure at sea level. The concept of atmospheric pressure is explained on the facing page.

Atmospheric Pressure (A.P.)

Atmospheric pressure (A.P.) is an important measurement. It's the collective force exerted by the mixture of all the gases in the atmosphere. At sea level, atmospheric pressure is equal to 760 mm Hg. As you ascend, atmospheric pressure decreases. Any time air comes in contact with any surface, atmospheric pressure comes into play. For example, consider the child below. Air comes in contact with every part of the body surface such as the head, shoulders, arms, hands, and all the rest.

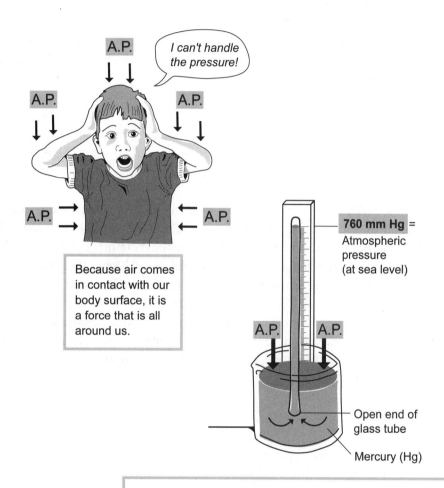

A.P.

I can't handle the pressure!

A.P.   A.P.

A.P.   A.P.

Because air comes in contact with our body surface, it is a force that is all around us.

760 mm Hg = Atmospheric pressure (at sea level)

A.P.   A.P.

Open end of glass tube

Mercury (Hg)

Model of a Barometer

A barometer measures atmospheric pressure (A.P.). When you listen to a weather report, you hear about changes in barometric pressure or A.P. The barometer above is a container filled with liquid mercury (Hg) that has an open-ended glass tube placed in it. As the A.P. presses on the surface of this liquid, it pushes the mercury into the open end of the glass tube. The force is measured as the height of the column in millimeters of mercury (mm Hg) with the scale on the side.

Concept: Pressure and Volume Law (Boyle's Law)

The pressure and volume law, termed **Boyle's law**, is a gas law; it states that **volume (V)** and **pressure (P)** have an inversely proportional relationship when the temperature and number of gas molecules remain constant. The illustration on the facing page shows a hollow sphere containing 5 gas molecules such as oxygen. These molecules have kinetic energy, so they randomly strike against the wall of the sphere. This is the source of the gas pressure inside the sphere.

Consider what will happen if the small sphere increases in size. Does the number of gas molecules change? No, 5 gas molecules are still present. The only thing that will change is the volume of the sphere. Well, if the volume increases, what do you predict will happen to the pressure inside? That's right—it will decrease.

Let's express this concept in a mathematical formula. If we refer to the volume and pressure in the small sphere as V_1 and P_1 and use V_2 and P_2 for the larger sphere, the pressure and volume law can be expressed as this equation:

$$V_1 \times P_1 = V_2 \times P_2$$

The product of the volume and the pressure in the small sphere should be equivalent to the product of the volume and the pressure in the larger sphere. Using the numbers given on the facing page, fill in the blanks to see if this relationship holds true.

Application: Mechanics of Breathing

Normal breathing is a repeated cycle of **inhalations** and **exhalations** and is an application of the pressure and volume law. At rest, the normal value of the pressure inside the lungs is equal to the atmospheric pressure outside the body, **1 atmosphere** (atm) or **760** mm Hg. The first step in a normal inhalation is contraction of a large muscle below the lungs called the **diaphragm**. As it contracts, it moves downward, increasing the volume in the lungs. This increase in volume causes a decrease in lung pressure to about 758 mm Hg. Air then moves down its pressure gradient and fills the lungs. In an exhalation, the diaphragm relaxes and moves upward, decreasing the volume in the lungs. This causes a slight increase in lung pressure to about 762 mm Hg. Air then rushes down its pressure gradient and flows out of the lungs. As you can see, normal breathing is a good application of the pressure and volume law.

CONCEPT: PRESSURE and VOLUME LAW (Boyle's Law)

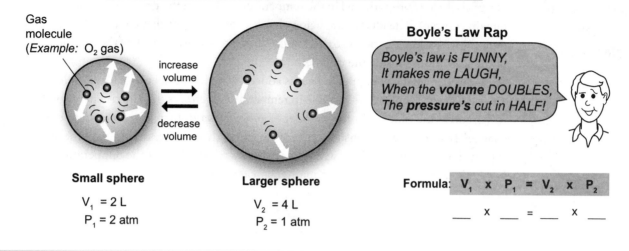

Gas molecule (*Example:* O_2 gas)

increase volume

decrease volume

Small sphere

V_1 = 2 L
P_1 = 2 atm

Larger sphere

V_2 = 4 L
P_2 = 1 atm

Boyle's Law Rap

Boyle's law is FUNNY,
It makes me LAUGH,
*When the **volume** DOUBLES,*
*The **pressure's** cut in HALF!*

Formula: $V_1 \times P_1 = V_2 \times P_2$

___ x ___ = ___ x ___

APPLICATION: MECHANICS of BREATHING

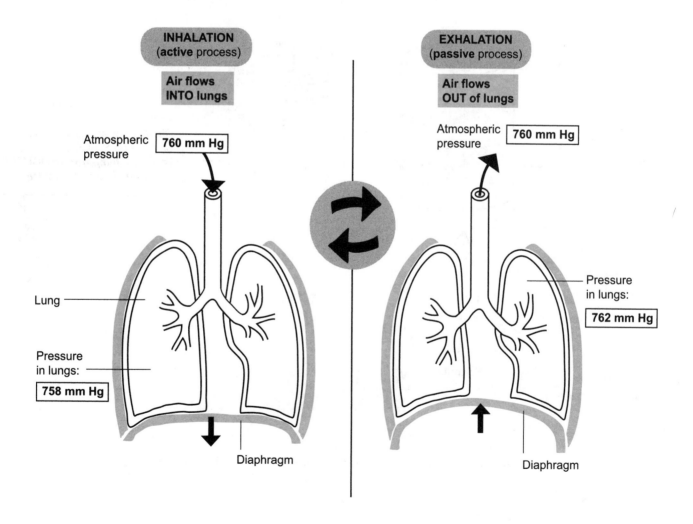

INHALATION (active process)

Air flows INTO lungs

Atmospheric pressure — 760 mm Hg

Lung

Pressure in lungs: 758 mm Hg

Diaphragm

EXHALATION (passive process)

Air flows OUT of lungs

Atmospheric pressure — 760 mm Hg

Pressure in lungs: 762 mm Hg

Diaphragm

Concept: Temperature and Volume Law (Charles' law)

The temperature and volume law, termed **Charles' law,** is a gas law; it states that the volume (V) of a gas is directly proportional to the temperature (T) when the pressure and number of gas molecules remain constant. The illustration below shows a clear cylinder that contains five gas molecules. These molecules have kinetic energy so they randomly strike against the wall of the cylinder. This is the source of the gas pressure inside the cylinder.

Consider what will happen if heat is added. Will the number of gas molecules change? No, there are still 5 gas molecules. The increase in temperature increases the kinetic energy of the gas molecules, which forces the plunger in the cylinder to move upward, increasing the volume. Does the pressure change? No, it remains constant. The only thing that changes is the volume in the cylinder. As expected, the volume doubles, which corresponds to doubling of the temperature.

Let's express this concept in a mathematical formula:

$$\frac{V_1}{T_1} = \frac{V_2}{T_2}$$

Using the numbers given in the illustration, fill in the blanks below for the mathematical formula.

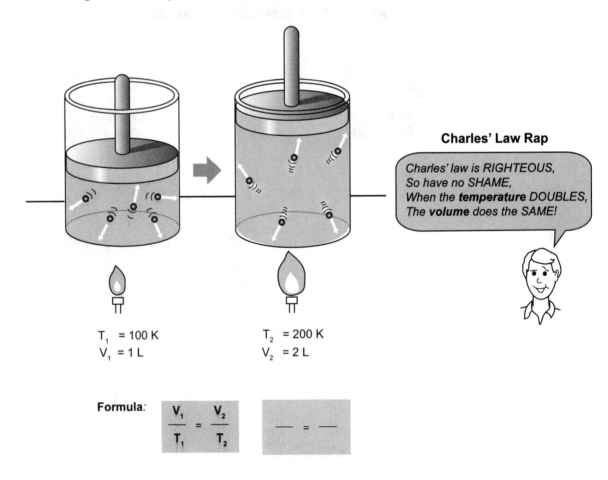

Charles' Law Rap

Charles' law is RIGHTEOUS,
So have no SHAME,
*When the **temperature** DOUBLES,*
*The **volume** does the SAME!*

$T_1 = 100 \text{ K}$
$V_1 = 1 \text{ L}$

$T_2 = 200 \text{ K}$
$V_2 = 2 \text{ L}$

Formula: $\quad \dfrac{V_1}{T_1} = \dfrac{V_2}{T_2} \qquad \underline{} = \underline{}$

On the facing page, we will examine two different applications of this gas law.

Application #1: Hot air ballooning

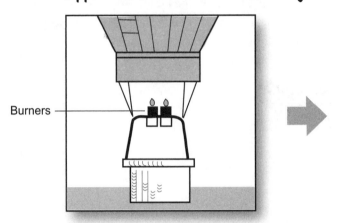

Burners

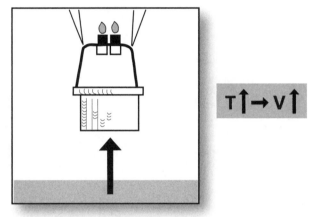

$T\uparrow \rightarrow V\uparrow$

Turning on the burners heats the air inside the balloon. This causes the volume to increase as the wall of the balloon stretches.

Because the hot air inside the balloon is less dense than the air outside, this creates lift.

Application #2: Mylar balloon shrinking

$T\downarrow \rightarrow V\downarrow$

Inside your home, at room temperature, the volume of the balloon remains constant.

When you take the balloon outside, on a snowy, winter day, the temperature decreases, causing the volume to also decrease.

Concept: Temperature and Pressure Law (Gay-Lussac's Law)

The **temperature and pressure law**, termed **Gay–Lussac's law**, is a gas law; it states that pressure (P) is directly proportional to the temperature (T) in Kelvin (K) when the volume and the number of gas molecules remain constant. In the illustration below, the clear cylinder contains 5 gas molecules. These molecules have kinetic energy, so they randomly strike the walls of the cylinder. This is the source of gas pressure inside the cylinder.

Consider what will happen if heat is added. Will the number of gas molecules change? No, there are still 5 gas molecules. The increase in temperature increases the kinetic energy of the gas molecules, which causes them to strike the container more frequently. This results in an increase in pressure (because there is no change in volume).

Let's express this concept in a mathematical formula:

$$\frac{P_1}{T_1} = \frac{P_2}{T_2}$$

Using the numbers given in the illustration, fill in the blanks below for the mathematical formula.

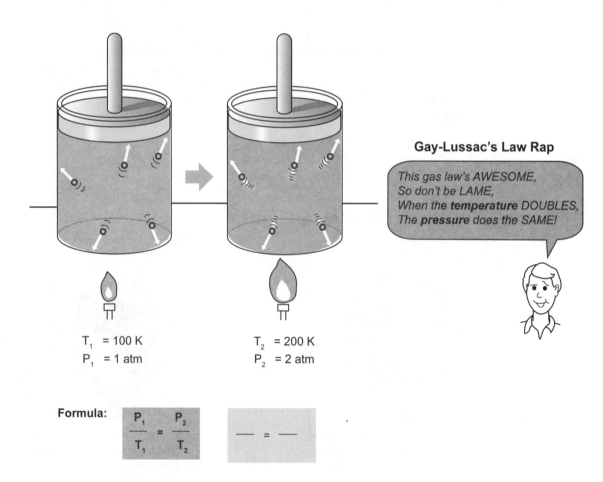

Gay-Lussac's Law Rap

This gas law's AWESOME,
So don't be LAME,
*When the **temperature** DOUBLES,*
*The **pressure** does the SAME!*

T_1 = 100 K
P_1 = 1 atm

T_2 = 200 K
P_2 = 2 atm

Formula: $\frac{P_1}{T_1} = \frac{P_2}{T_2}$ $\underline{\quad} = \underline{\quad}$

On the facing page, we will examine two different applications of this gas law.

Application #1: Aerosal can being heated

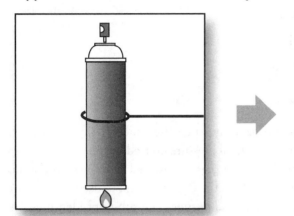

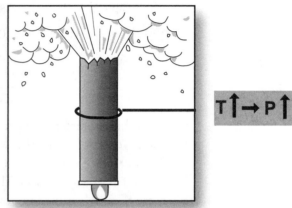

Aerosal cans have a warning label stating that they should not be heated. If a can is heated . . .

. . . the temperature of the gas inside the can increases proportionally with the pressure. When the pressure becomes too great, the can explodes! (*Note:* Do not try this at home.)

Application #2: Pressure cooker

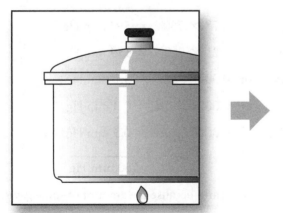

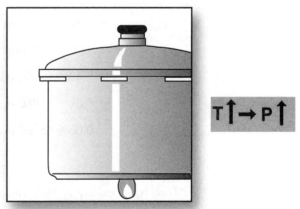

Pressure cookers use steam at high temperatures to cook food. Note that the volume of the container remains constant.

The temperature of the steam increases proportionally with the pressure. This has the benefit of cooking food quickly.

Description

The **law of partial pressures** of gases, termed **Dalton's law**, states that in a mixture of gases (such as the air we breathe), each gas exerts a pressure referred to as a "partial pressure." Therefore, the sum of all the partial pressures is the total pressure exerted by the gas mixture. Air contains the following gases:

Nitrogen (N_2)	78.0 %
Oxygen (O_2)	21.0 %
Water vapor (H_2O)	0.40 %
Carbon dioxide (CO_2)	0.04 %
Other gases	0.06 %
	100%*

Consider the barometer on the facing page. It measures the atmospheric pressure, which at sea level is equal to 760 mm Hg. In other words, the total pressure exerted by the gas mixture in air is 760 mm Hg. But what if we want to calculate the partial pressure exerted by *each gas* in the mixture? First, we have to understand symbolism. The letter "P" is the chemical symbol used for pressure, and "P" with a subscript, or "P_x", is used to designate a *partial pressure*. The subscript identifies the given gas. For example, here is how the partial pressures exerted by nitrogen and oxygen would be written:

P_{N_2} = the partial pressure exerted by nitrogen

P_{O_2} = the partial pressure exerted by oxygen

The force each gas exerts is proportional to its percentage in the atmosphere. To calculate, simply multiply the percentage of the gas in the mixture by the total pressure exerted by the gas mixture as detailed below.

P_{N_2}	= 78% of 760 mm Hg	or 0.78 × 760	= **597** mm Hg
P_{O_2}	= 21% of 760 mm Hg	or 0.21 × 760	= **159** mm Hg
P_{H_2O}	= 0.4% of 760 mm Hg	or 0.004 × 760	= **3.2** mm Hg
P_{CO_2}	= 0.04% of 760 mm Hg	or 0.0004 × 760	= **0.3** mm Hg
$P_{other\ gases}$	= 0.06% of 760 mm Hg	or 0.0006 × 760	= **0.5** mm Hg
			760 mm Hg

*Total is not exactly 100% because of rounding and because the amount of some gases, such as water vapor, varies with the environment.

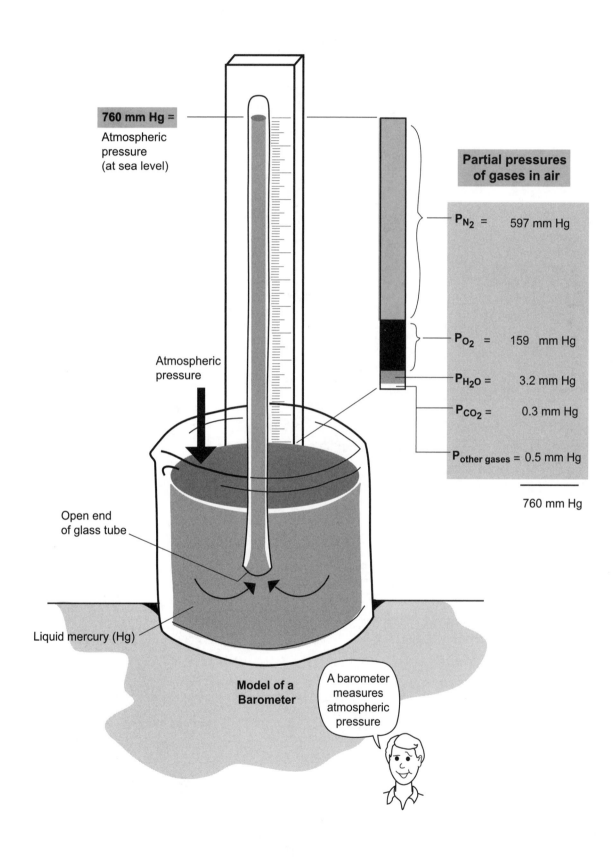

760 mm Hg =
Atmospheric
pressure
(at sea level)

Atmospheric
pressure

Open end
of glass tube

Liquid mercury (Hg)

**Model of a
Barometer**

A barometer
measures
atmospheric
pressure

**Partial pressures
of gases in air**

P_{N_2} =   597 mm Hg

P_{O_2} =   159   mm Hg

P_{H_2O} =   3.2 mm Hg

P_{CO_2} =   0.3 mm Hg

$P_{\text{other gases}}$ = 0.5 mm Hg

760 mm Hg

Solutions

Description

A **solution** is a homogeneous mixture consisting of a substance (solute) dissolved and uniformly distributed in another (solvent). A common example is a **saline** (salt) **solution**. Water is the **solvent** that dissolves sodium chloride (NaCl), the **solute**.

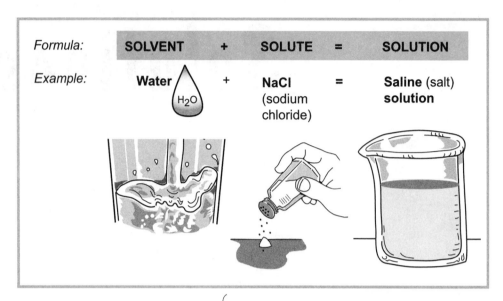

Formula: **SOLVENT + SOLUTE = SOLUTION**

Example: **Water** + **NaCl** (sodium chloride) = **Saline** (salt) **solution**

Liquid Solutions

Water serves as the solvent for most **liquid solutions**. These solutions are called aqueous. Solutes can be gases (*Example*: CO_2), solids (*Example*: NaCl), and/or liquids (*Example*: acetic acid) dissolved in water. Wine, vinegar, and seawater are all examples of liquid solutions.

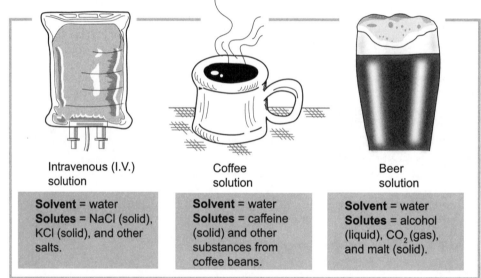

Intravenous (I.V.) solution

Solvent = water
Solutes = NaCl (solid), KCl (solid), and other salts.

Coffee solution

Solvent = water
Solutes = caffeine (solid) and other substances from coffee beans.

Beer solution

Solvent = water
Solutes = alcohol (liquid), CO_2 (gas), and malt (solid).

Gas and Solid Solutions

In additon to liquid solutions, there are also **gas solutions** (the solvent and the solute are both gases) and **solid solutions** (the solvent and the solute are both solids). The solvent is the substance present in a greater amount and surrounds the solute.

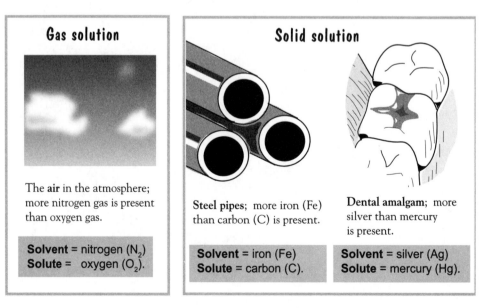

Gas solution

The **air** in the atmosphere; more nitrogen gas is present than oxygen gas.

Solvent = nitrogen (N_2)
Solute = oxygen (O_2).

Solid solution

Steel pipes; more iron (Fe) than carbon (C) is present.

Solvent = iron (Fe)
Solute = carbon (C).

Dental amalgam; more silver than mercury is present.

Solvent = silver (Ag)
Solute = mercury (Hg).

Water as a Solvent

Water is a common solvent. Each water molecule (H_2O) contains two hydrogens (H) covalently bonded to one oxygen (O) atom. Like a magnet, the hydrogen ends of each O–H bond possess a partially positive charge ($\delta +$) and the oxygen end has a partially negative charge ($\delta -$). This partial charge makes these bonds polar. The slightly positive H of one molecule is attracted to the slightly negative O of another molecule to form a hydrogen bond. Because of water's shape and its polar O–H bonds, water is a polar molecule.

The chemical rule for dissolving is: "like dissolves like." Water is polar, so it dissolves other polar compounds but not nonpolar compounds such as corn oil. When crystals of NaCl are added to water, the water molecules are attracted to its charge. The oxygen ends are attracted to the sodium ions (Na^+), and the hydrogen ends are attracted to the chloride ions (Cl^-). The final result is that the ionic bond between NaCl salt is broken, the ions are separated, and water molecules surround each ion. This action (solvation) is what you observe as you add sodium chloride crystals to water, and they seem to "disappear" as you stir them into solution.

Water

Hydrogen bond

Water molecule (H_2O)

Sodium chloride solution

Practice Problems

You make a sugar solution to put in your hummingbird feeder. Answer the following:

1 What is the **solvent** in your sugar solution?

2 What is the **solute** in your sugar solution?

3 What type of solution did you make? Explain.

Answers

1 water 2 sugar 3 It's an aqueous solution, a type of liquid solution because the solid solute (sugar) was dissolved in a liquid solvent (water).

SOLUTIONS

Solution components (electrolytes and nonelectrolytes)

ELECTROLYTES

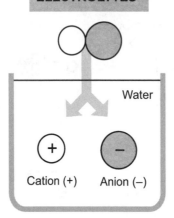

Water

(+) Cation (+)

(−) Anion (−)

Definition:

An **electrolyte** is any substance that dissolves and ionizes in solution and thereby conducts electricity.

Examples:

Compounds such as:

- **Salts** – NaCl, CaCl$_2$
- **Acids** – HCl, CH$_3$COOH (acetic acid)
- **Bases** – NaOH, NH$_3$ (ammonia)

STRONG electrolytes

Strong electrolytes *completely* ionize in solution and are very good conductors of electricity.

Examples:

- **Salts** – NaCl, KCl
- **Strong acids** – HCl, HNO$_3$
- **Strong bases** – NaOH, KOH

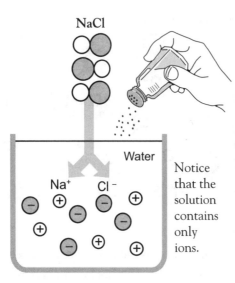

NaCl

Water

Na$^+$ Cl$^-$

Notice that the solution contains only ions.

Table salt (sodium chloride, **NaCl**) is a strong electrolyte

$$NaCl \xrightarrow{H_2O} Na^+ + Cl^-$$

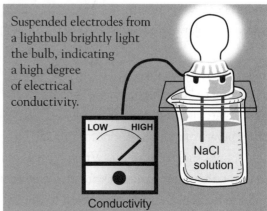

Suspended electrodes from a lightbulb brightly light the bulb, indicating a high degree of electrical conductivity.

LOW   HIGH

NaCl solution

Conductivity

WEAK electrolytes

Weak electrolytes only *partially* ionize in solution and are poor conductors of electricity.

Examples:

- **Weak acids** – CH$_3$COOH (acetic acid)
- **Weak bases** – NH$_3$ (ammonia)

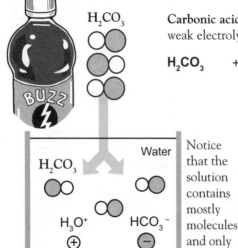

H$_2$CO$_3$

Water

H$_2$CO$_3$

H$_3$O$^+$ HCO$_3$$^-$

Notice that the solution contains mostly molecules and only some ions.

Carbonic acid (**H$_2$CO$_3$**) gives soda pop its bubbles and is a weak electrolyte

$$H_2CO_3 + H_2O \rightleftharpoons H_3O^+ + HCO_3^-$$

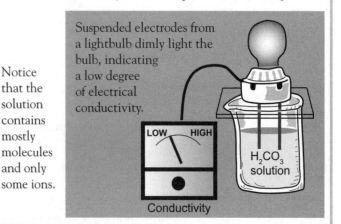

Suspended electrodes from a lightbulb dimly light the bulb, indicating a low degree of electrical conductivity.

LOW   HIGH

H$_2$CO$_3$ solution

Conductivity

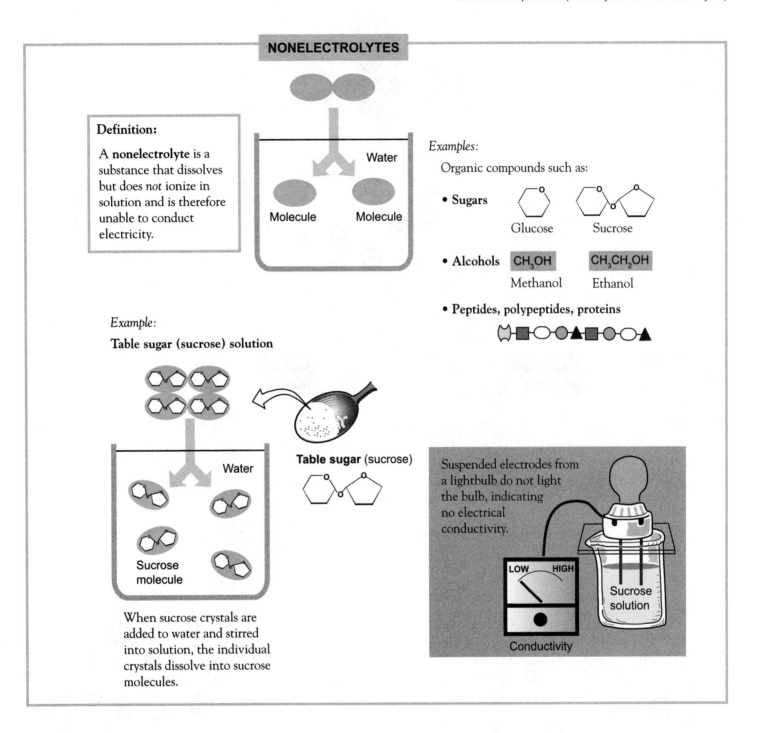

NONELECTROLYTES

Definition:

A **nonelectrolyte** is a substance that dissolves but does *not* ionize in solution and is therefore unable to conduct electricity.

Water

Molecule   Molecule

Examples:

Organic compounds such as:

- **Sugars**

  Glucose   Sucrose

- **Alcohols** CH₃OH CH₃CH₂OH

  Methanol   Ethanol

- **Peptides, polypeptides, proteins**

Example:

Table sugar (sucrose) solution

Table sugar (sucrose)

Water

Sucrose molecule

When sucrose crystals are added to water and stirred into solution, the individual crystals dissolve into sucrose molecules.

Suspended electrodes from a lightbulb do not light the bulb, indicating no electrical conductivity.

LOW   HIGH

Sucrose solution

Conductivity

Practice Problems

Determine if the following solutions contain strong electrolytes, weak electrolytes, or nonelectrolytes:

1   Insulin (protein) solution

2   Intravenous solution containing NaCl and KCl

3   Vinegar solution

4   Cleaning solution with ammonia

Answers

Definition and Examples

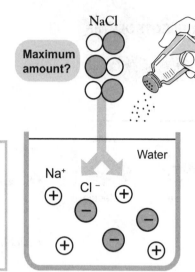

Definition:

Solubility is the maximum amount of solute that can be dissolved in a solvent at a given temperature.

Examples:

Solubility is expressed as grams of solute per 100 grams of water at a **given temperature**.

For **NaCl** at 20°C: 36 g/100 g of H_2O

For **NaNO$_3$** at 20°C: 81 g/100 g of H_2O

For **CaBr$_2$** at 20°C: 144 g/100 g of H_2O

Unsaturated and Saturated Solutions

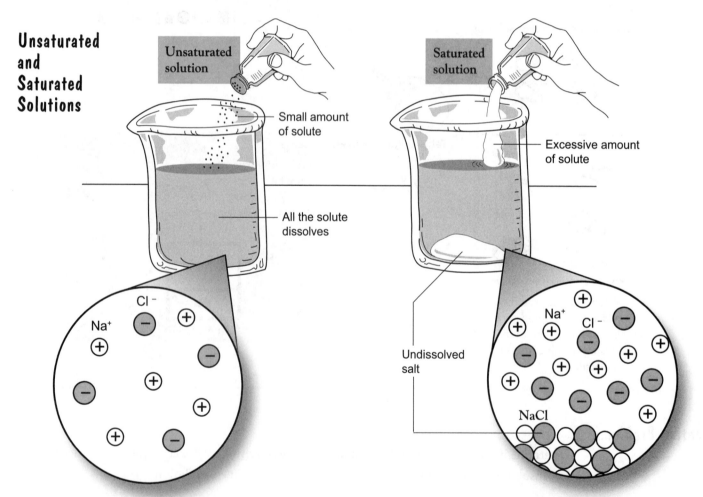

Unsaturated solution — Small amount of solute — All the solute dissolves

Saturated solution — Excessive amount of solute — Undissolved salt

Unsaturated solutions contain **less** solute than can be solubilized by the solvent. The evidence for this is that all the salt dissolves.

Saturated solutions contain **more** solute than can be solubilized by the solvent. The evidence for this is the presence of undissolved salt.

Practice Problems

1 You mix an unweighed sample of a solid in 100 mL of water. After stirring, the mixture is transparent and no solid is visible. What is the mixture called?

2 You mix an unweighed sample of a solid in 100 mL of water. After stirring, the mixture is transparent but much of the solid rests on the bottom of the beaker. What is this mixture called?

3 You mix 120 g of calcium bromide ($CaBr_2$) in 100 mL of water at 20°C. Is this solution saturated or unsaturated? Explain.

Answers

1 An unsaturated solution   2 A saturated solution

3 This solution is unsaturated because 120 g is less than the solubility factor of 144 g at 20°C.

Definition and Properties

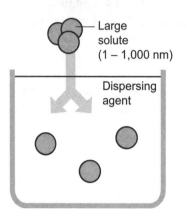

COLLOIDS

Large solute
(1 – 1,000 nm)

Dispersing agent

Definition:

A **colloid** is a heterogeneous mixture containing either large solute particles or solid, liquid, or gas aggregates scattered within a dispersing agent such as a gas, a liquid, or a solid.

Properties of a colloid:

- **Particle size** – solute particle is large, roughly 1 to 1,000 nm

- **Turbidity** – scatters light so it appears turbid or cloudy

- **Settling** – solute particles do **NOT** settle

- **Separation** – solute particles **CANNOT** be separated by filters

General Examples of Colloids

Colloids can be found in the kitchen and other common places. It's important to note that they can involve substances in different states such as a gas in a liquid or a gas in a solid.

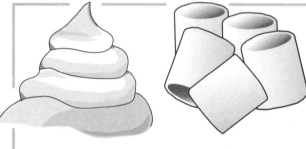

Whipped cream

Gas in a liquid.

(nitrous oxide [N_2O] gas is dispersed in liquid cream)

Marshmallows

Gas in a solid.

(air is dispersed in a solid, gelatinous sugar mixture)

Milk

Liquid in a liquid.

(lipid globules are dispersed in water)

Biological Examples

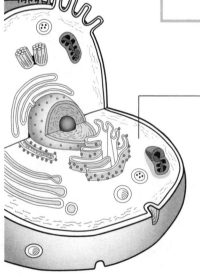

Body cell

The **cytosol** is the liquid portion of the cell (excluding the organelles) and it is both a solution and a colloid. It's a solution because it contains many different ions dissolved in water. But it also contains large organic substances like proteins that are dispersed in the water, which makes it a colloid.

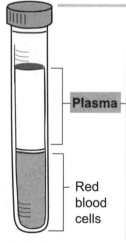

Plasma

Red blood cells

Tube of centrifuged blood

The **plasma** is an example of a colloid. When whole blood is spun in a centrifuge it separates into its liquid portion—the plasma—and its solid portion—mostly red blood cells. Various proteins are scattered in water, which gives the plasma a pale yellow color.

SUSPENSIONS

Definition and Properties

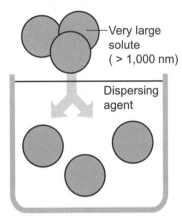

Very large solute
(> 1,000 nm)

Dispersing agent

Definition:

A **suspension** is a mixture containing very large solute particles evenly distributed in a dispersing agent by mechanical means that will eventually settle out.

Properties of a suspension:

- **Particle size** – solute particle is very large, > 1,000 nm

- **Turbidity** – scatters light so it appears very turbid or cloudy

- **Settling** – solute particles settle quickly

- **Separation** – solute particles can be separated by filters

General Examples of Suspensions

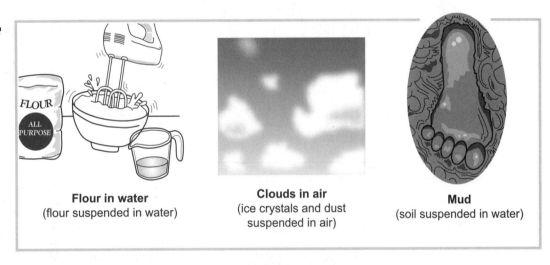

Flour in water
(flour suspended in water)

Clouds in air
(ice crystals and dust suspended in air)

Mud
(soil suspended in water)

Biological Examples

Cytoplasm is composed of the liquid portion inside the cell called the cytosol and all the very large organelles such as the mitochondrion. These organelles are suspended in the cytosol, making the cytoplasm a suspension.

Whole blood is a suspension because red and white blood cells are suspended in the liquid portion of the blood called the plasma. Cell fragments called platelets are another large solute in this suspension.

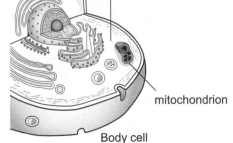

mitochondrion

Body cell

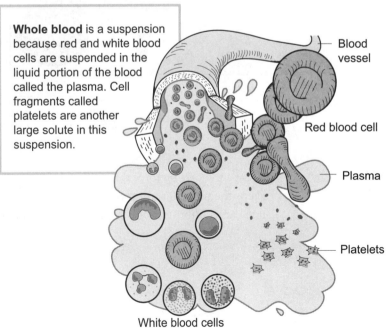

Blood vessel

Red blood cell

Plasma

Platelets

White blood cells

Definition and Demonstration

BEFORE

A dialysis bag containing a solution of glucose (small solute) and protein (large solute) is placed in a beaker of water. The protein is too big to pass through the bag, and the glucose is small enough to pass through.

AFTER

After a short period of time, the glucose moves out of the bag and into the water until the glucose concentrations inside and outside the bag are equal.

Definition:

Dialysis is the separation of large solutes from small solutes by allowing the latter to pass through a semi-permeable membrane.

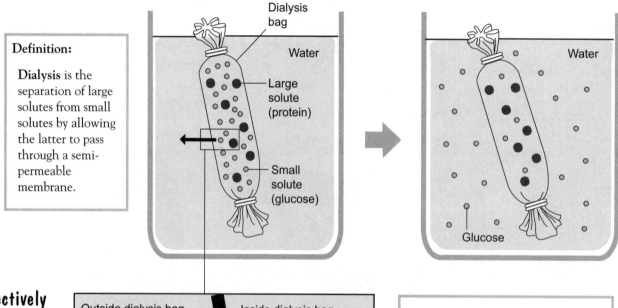

Selectively Permeable Membrane

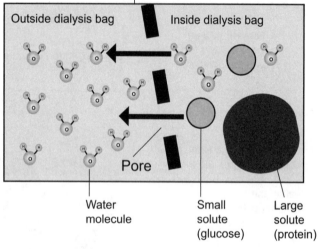

The wall of the dialysis bag has microscopic pores in it. This allows both small solute molecules (glucose) and solvent molecules (water) to pass through the membrane, but not the large solute molecules. In other words, the wall of the dialysis bag is **selectively permeable** to particles based on their size. The same is true of your body's cell membranes.

Driving Force?

What's the driving force that makes dialysis occur? The kinetic energy of the small molecules is what keeps them in constant motion and drives them through the pores in the semipermeable membrane until an equlibrium state is reached across the membrane.

Application: Hemodialysis (Kidney Dialysis)

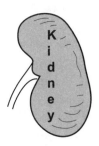

The function of the kidneys is to remove urea and other toxic wastes from the blood, to produce urine, and to deliver the urine to the urinary bladder. A person who has permanent kidney damage requires **hemodialysis** (kidney dialysis) performed on a regular basis to remove these waste products. A **dialyzer** functions as a simple artificial kidney. It contains a special semipermeable tubing—**dialysis tubing**—through which the patient's blood passes. The tubing is surrounded by a **dialyzing solution**—a saline solution with similar components to the blood plasma such as $NaCl$, KCl, and $NaHCO_3$. As urea and other waste products move from the blood and into the dialyzing solution, the urea concentration increases in the dialyzing solution. Used dialyzing solution containing urea is removed and replaced regularly with fresh dialysis solution. This step keeps urea moving out of the blood by preventing an equilibrium state from occurring between the urea concentration in the blood and the dialyzing solution.

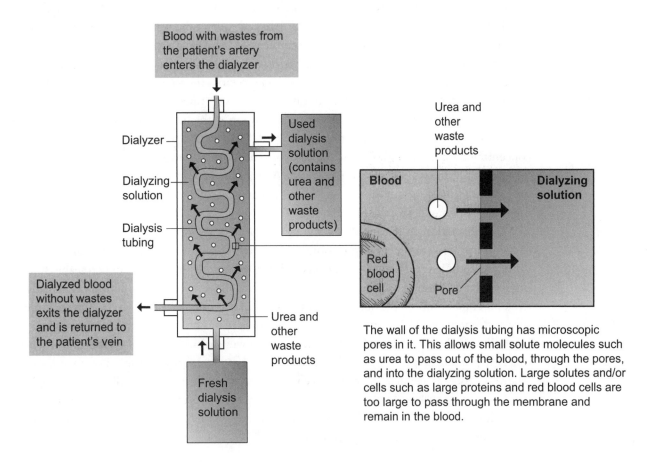

The wall of the dialysis tubing has microscopic pores in it. This allows small solute molecules such as urea to pass out of the blood, through the pores, and into the dialyzing solution. Large solutes and/or cells such as large proteins and red blood cells are too large to pass through the membrane and remain in the blood.

Description

In biological systems, **osmosis** is defined as the flow of **water** across a semipermeable membrane toward the solution with the higher solute concentration. This common, important process occurs when water moves in and out of plant and animal cells by crossing their cell membranes. The driving force is the tendency toward achieving dynamic equilibrium, in which there is an equal concentration of solute on both sides of the cell membrane. When examining osmosis, the solute in question is impermeable to the cell membrane. It is also essential to realize that the solution inside the cell (intracellular solution) is always compared to the solution outside the cell (extracellular solution). For example, if you told me I was tall, I might ask, "Compared to what?" Compared to a toddler, I may be tall but not when compared to a giraffe. Similarly, if you were to tell me that a solution was concentrated, I could also ask, "Compared to what?" The word "concentrated" makes sense only in the context of a relative comparison. While your body has regulatory mechanisms to keep the concentration of the intracellular and extracellular solutions relatively stable, the extracellular solution is more likely to change in its solute concentration due to external influences. Sometimes it is more dilute, other times it is more concentrated.

The illustration of the hollow, glass, "U"-shaped tube at the top of the facing page demonstrates the process of osmosis. An artificial, semipermeable membrane separates two solutions of different concentrations—one on side A and the other on side B. Notice that the solute in both solutions is a protein that is impermeable because it is too large to cross the membrane. The solution on side B is more concentrated than the solution on side A. Any time the two solutions are unequal in concentration, water always moves toward the solution with the higher solute concentration. In this case, water will move from side A to side B. Why? Think of this as water's attempt to dilute the more concentrated solution to reach dynamic equilibrium. As water moves, the volume of solution on side B increases while the volume of solution on side A decreases. Water stops moving when the two solutions have equal solute concentrations—the state of dynamic equilibrium. At dynamic equilibrium, the net movement of water across the membrane is zero.

The illustration in the middle of the facing page shows a red blood cell in three different types of solutions: (1) **isotonic**, (2) **hypertonic**, and (3) **hypotonic**. Your body's extracellular solutions are normally **isotonic** (iso = equal) **solutions** with the same concentration of impermeable solutes as the intracellular solution. This is a stable state for body cells. A cell in this solution has already achieved dynamic equilibrium, so the net movement of water in and out of the cell is zero. A **hypertonic solution** (hyper = more, greater) is one that has a greater concentration of impermeable solutes relative to the intracellular solution. A cell placed in this solution will shrink (crenate) because of water leaving the cell. A **hypotonic** (hypo = less) **solution** is one that has a lesser concentration of impermeable solutes relative to the intracellular solution. A cell placed in this solution will swell and possibly burst (lyse) because of water rushing into the cell. By convention, biologists typically use these terms to describe the extracellular solution, though they technically can be used to describe either the intracellular or the extracellular solution as a relative comparison.

Applications

Three applications for osmosis are illustrated on the bottom of the facing page:

1. Using an isotonic solution is illustrated by giving a patient an intravenous (I.V.) saline solution at the hospital. The saline solution has to be isotonic to your red blood cells so they neither shrink nor burst.

2. Creating a hypertonic solution is illustrated by adding salt to the slimy film coating the surface of the slug. As a result, water leaves the cells of the slug and can cause it to shrivel up and die.

3. Using a hypotonic solution is illustrated by placing wilted lettuce in a bowl of water in the refrigerator overnight to refresh it. Water in the bowl enters the cells of the lettuce leaf, making it firm again.

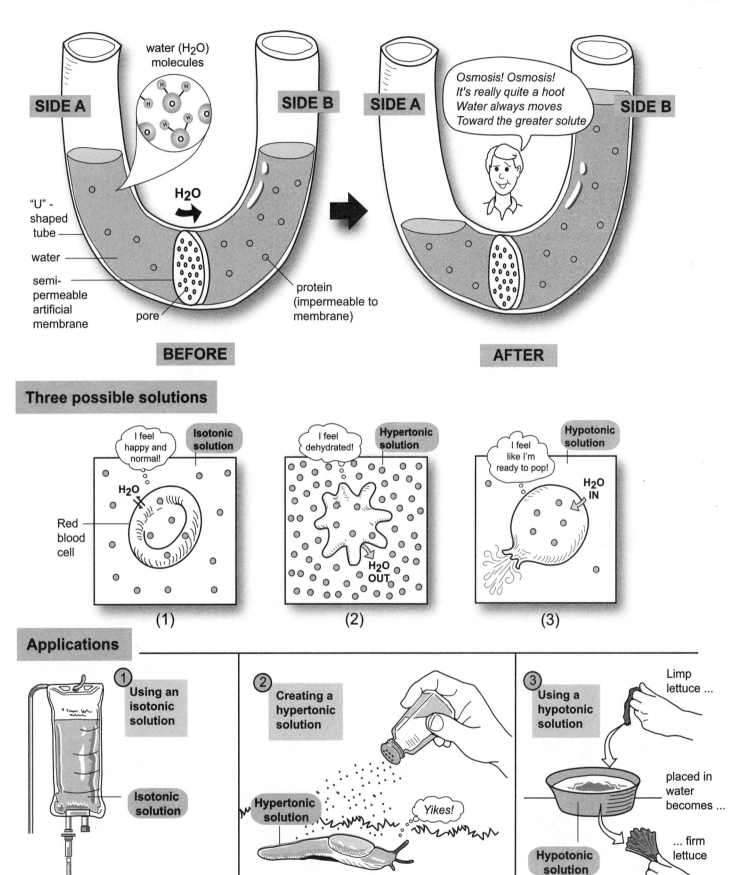

SIDE A — water (H₂O) molecules — SIDE B

"U"-shaped tube

water

semi-permeable artificial membrane

pore

H₂O

protein (impermeable to membrane)

BEFORE

SIDE A — SIDE B

*Osmosis! Osmosis!
It's really quite a hoot
Water always moves
Toward the greater solute*

AFTER

Three possible solutions

(1) I feel happy and normal! — Isotonic solution — H₂O — Red blood cell

(2) I feel dehydrated! — Hypertonic solution — H₂O OUT

(3) I feel like I'm ready to pop! — Hypotonic solution — H₂O IN

Applications

1 Using an isotonic solution — Isotonic solution

2 Creating a hypertonic solution — Hypertonic solution — Yikes!

3 Using a hypotonic solution — Hypotonic solution — Limp lettuce ... placed in water becomes ... ... firm lettuce

109

Acids
and Bases

ACIDS

Definition and Common Substances Containing Acids

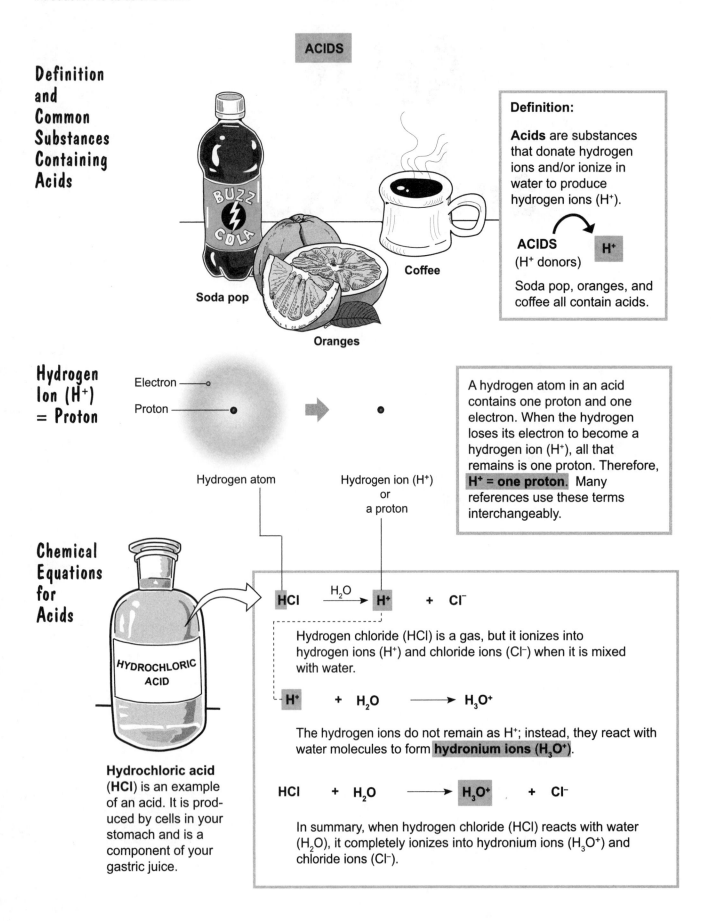

Soda pop

Oranges

Coffee

Definition:

Acids are substances that donate hydrogen ions and/or ionize in water to produce hydrogen ions (H^+).

ACIDS
(H^+ donors) → H^+

Soda pop, oranges, and coffee all contain acids.

Hydrogen Ion (H^+) = Proton

Electron

Proton

Hydrogen atom

Hydrogen ion (H^+)
or
a proton

A hydrogen atom in an acid contains one proton and one electron. When the hydrogen loses its electron to become a hydrogen ion (H^+), all that remains is one proton. Therefore, H^+ = **one proton**. Many references use these terms interchangeably.

Chemical Equations for Acids

HYDROCHLORIC ACID

Hydrochloric acid (HCl) is an example of an acid. It is produced by cells in your stomach and is a component of your gastric juice.

$$HCl \xrightarrow{H_2O} H^+ + Cl^-$$

Hydrogen chloride (HCl) is a gas, but it ionizes into hydrogen ions (H^+) and chloride ions (Cl^-) when it is mixed with water.

$$H^+ + H_2O \longrightarrow H_3O^+$$

The hydrogen ions do not remain as H^+; instead, they react with water molecules to form **hydronium ions (H_3O^+)**.

$$HCl + H_2O \longrightarrow H_3O^+ + Cl^-$$

In summary, when hydrogen chloride (HCl) reacts with water (H_2O), it completely ionizes into hydronium ions (H_3O^+) and chloride ions (Cl^-).

BASES

Definition and Common Substances Containing Bases

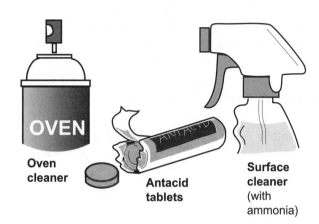

Oven cleaner

Antacid tablets

Surface cleaner (with ammonia)

Definition:

Bases are substances that accept hydrogen ions and/or ionize in water to produce hydroxide ions (OH^-).

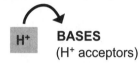

H^+ **BASES** (H^+ acceptors)

Oven cleaner, antacid tablets, and any surface cleaner containing ammonia all contain bases.

Chemical Equations for Bases

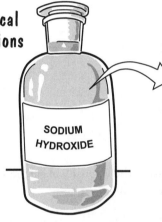

SODIUM HYDROXIDE

Sodium hydroxide (NaOH) is an example of a base that is a metal hydroxide.

$$NaOH \xrightarrow{H_2O} Na^+ + OH^-$$

Sodium hydroxide (NaOH) is a solid, but it ionizes to form sodium ions and hydroxide ions (OH^-) when it is mixed with water. The H_2O above the arrow indicates that water is involved in the ionization process but does not serve as a chemical reactant.

Many bases are metal hydroxides such as NaOH, KOH, and $Mg(OH)_2$, but some are not.

Let's look at an example of another base that is not a metal hydroxide—namely, **ammonia (NH$_3$)**.

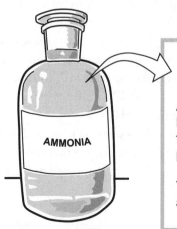

AMMONIA

Ammonia (NH$_3$) is an example of a base that is *not* a metal hydroxide.

$$NH_3 + H_2O \rightleftharpoons NH_4^+ + OH^-$$

Ammonia (NH$_3$) is a gas that forms ammonium ions (NH_4^+) and hydroxide ions (OH^-) when reacted with liquid water. NH_3 accepts H^+ from a water molecule to form NH_4^+. This leaves behind an OH^- in the process.

This perfectly fits the definition of a base as being both a substance that accepts hydrogen ions (H^+) and produces hydroxide ions (OH^-).

Description

The pH scale stands for "potential hydrogen" scale because it measures the molar concentration of hydrogen ions (H^+) in solution. Brackets [ ] are used to indicate molar concentration so the phrase "molar concentration of hydrogen ions in solution" can be abbreviated as [H^+]. But hydrogen ions immediately bind to water molecules to form hydronium ions (H_3O^+), so pH is really measuring [H_3O^+] in solution. To avoid confusion, because a hydrogen ion is a proton, it's important to note that all three of these are equivalent when discussing hydrogen ions in solution:

> **hydrogen ion (H^+) = proton = hydronium ion (H_3O^+)**

The pH values identify solutions as being acidic, or basic (alkaline), or neutral.

pH Scale Facts

Range:	0–14
Acid:	< 7.0
Base:	> 7.0
Neutral point:	7.0 – this is where [H_3O^+] = [OH^-].
Increments:	Each number represents a 10x (10-fold) change in [H_3O^+]
Units:	[H_3O^+] per liter of solution

Exponential (Logarithmic) Scale

① pH is mathematically calculated as: $pH = -\log[H_3O^+]$

② CpH is a logarithmic scale like the Richter scale used to measure the intensity of earthquakes. An earthquake at 6 is 1,000 times more intense than an earthquake measured at 3.

③ Changes in pH reflect exponential changes in the [H_3O^+] of the solution. Each unit on the scale represents a 10x (10-fold) change in [H_3O^+], so changes occur as 10^x, where x = the number of units changed on the scale.

For example, a solution with a pH of 5 is moderately acidic, like coffee. If we add acid and the pH decreases from 5 to 3, how much of a change is it? Well, $5-3 = 2$, so the solution changes 2 pH units. Therefore, $10^2 = 10 \times 10 = 100$, so there is a 100x change in [H_3O^+]. But is it an increase or a decrease in [H_3O^+]? If the pH value decreases, the solution is more acidic. Likewise, if the pH value of a solution increases, it is becoming more basic and therefore less acidic.

Now, if you add a base to your pH = 3 solution and the pH increases to 7, you are decreasing the [H_3O^+] by 4 pH units. Similarly, $3-7 = -4$, so $10^{-4} = 10^{-1} \times 10^{-1} \times 10^{-1} \times 10^{-1} = 1/10,000$. At a pH of 7, the solution has 1/10,000 the [H_3O^+] as it did at pH = 3.

So if you are hired as a lab technician and are asked to make a solution with a pH of 5 but you are a little sloppy and make a solution with a pH of 6 instead, you can see that you have made a big error. Your solution has a [H_3O^+] that is 10 times lower than it is supposed to be. Let's hope you don't get fired!

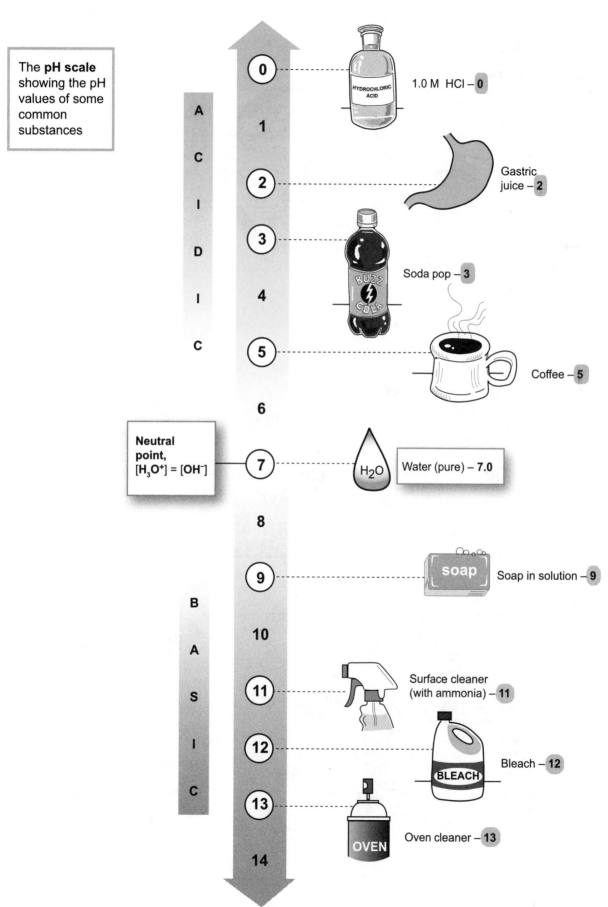

The **pH scale** showing the pH values of some common substances

ACIDIC

1.0 M HCl – **0**

Gastric juice – **2**

Soda pop – **3**

Coffee – **5**

Neutral point, $[H_3O^+] = [OH^-]$

Water (pure) – **7.0**

BASIC

Soap in solution – **9**

Surface cleaner (with ammonia) – **11**

Bleach – **12**

Oven cleaner – **13**

Water (pure)

At the molecular level, you might assume that pure water consists simply of water molecules (H_2O), but this is only partly correct because water molecules become ions.

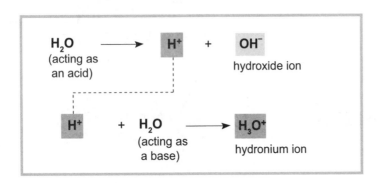

H_2O (acting as an acid) → H^+ + OH^- hydroxide ion

H^+ + H_2O (acting as a base) → H_3O^+ hydronium ion

What really happens is that some H_2O molecules act as an acid by donating a H^+ to a second H_2O acting as a base. The result is the formation of a hydronium (H_3O^+) ion.

H_2O + H_2O ⇌ H_3O^+ + OH^-

In summary, this equation best expresses the final result: When hydrogen ions are transferred between water molecules, the products formed are hydronium ions and hydroxide ions. The longer arrow indicates the greater tendency of the products to re-form into water molecules.

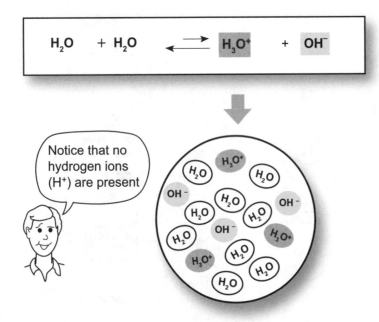

Notice that no hydrogen ions (H^+) are present

If we convert this equation into a graphic, it would look like this. In pure water, water molecules, hydronium ions, and hydroxide ions are all present. But water molecules are still present in the greatest amount.

Water Acting as an Acid and a Base

In pure water, some water molecules act as acids—hydrogen ion (H⁺) donors—and others act as bases—hydrogen ion (H⁺) acceptors.

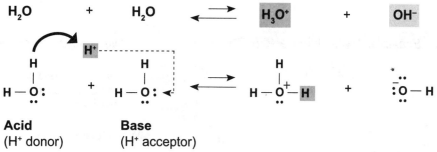

Acid
(H⁺ donor)

Base
(H⁺ acceptor)

The Relationship Between [H₃O⁺] and [OH⁻] for Every pH Value

[ ] = molar concentration

For **ACIDS**,
[H₃O⁺] > [OH⁻]

At 7.0, the **neutral point**,
[H₃O⁺] = [OH⁻]

For **BASES**,
[H₃O⁺] < [OH⁻]

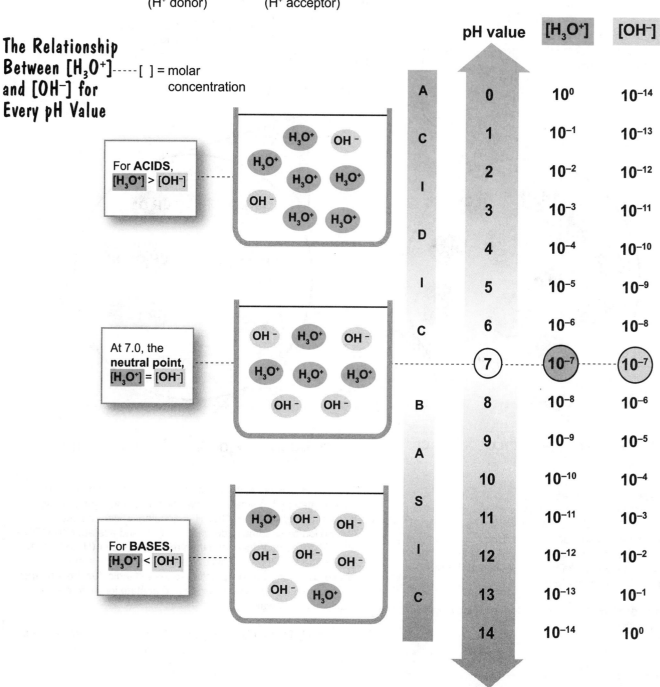

	pH value	[H₃O⁺]	[OH⁻]
A	0	10^0	10^{-14}
C	1	10^{-1}	10^{-13}
	2	10^{-2}	10^{-12}
I	3	10^{-3}	10^{-11}
D	4	10^{-4}	10^{-10}
I	5	10^{-5}	10^{-9}
C	6	10^{-6}	10^{-8}
	7	10^{-7}	10^{-7}
B	8	10^{-8}	10^{-6}
A	9	10^{-9}	10^{-5}
	10	10^{-10}	10^{-4}
S	11	10^{-11}	10^{-3}
I	12	10^{-12}	10^{-2}
C	13	10^{-13}	10^{-1}
	14	10^{-14}	10^0

ACIDS AND BASES

Strength of acids and bases

STRONG and WEAK ACIDS

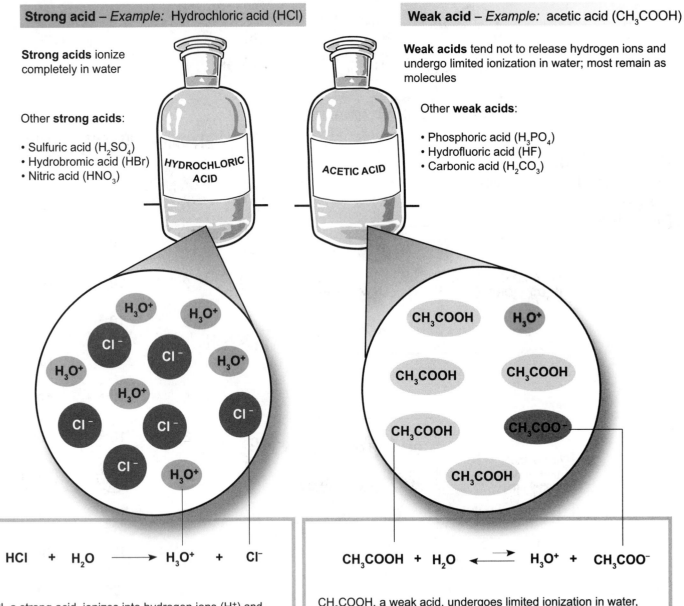

Strong acid – *Example:* Hydrochloric acid (HCl)

Strong acids ionize completely in water

Other **strong acids**:

• Sulfuric acid (H_2SO_4)
• Hydrobromic acid (HBr)
• Nitric acid (HNO_3)

Weak acid – *Example:* acetic acid (CH_3COOH)

Weak acids tend not to release hydrogen ions and undergo limited ionization in water; most remain as molecules

Other **weak acids**:

• Phosphoric acid (H_3PO_4)
• Hydrofluoric acid (HF)
• Carbonic acid (H_2CO_3)

$$HCl + H_2O \longrightarrow H_3O^+ + Cl^-$$

HCl, a strong acid, ionizes into hydrogen ions (H^+) and chloride (Cl^-) ions in water. The free hydrogen ions then react with water (H_2O) to form hydronium ions (H_3O^+). This explains the chemical equation shown above. The arrow pointing to the right indicates that HCl completely ionizes to form its component parts. Almost none of the HCl remains in its molecular form.

Strong acids follow this same pattern of completely ionizing in water.

$$CH_3COOH + H_2O \rightleftharpoons H_3O^+ + CH_3COO^-$$

CH_3COOH, a weak acid, undergoes limited ionization in water, as indicated by the smaller arrow pointing to the right. Instead, it remains mostly as CH_3COOH molecules. Those molecules that do ionize become hydrogen ions (H^+) and acetate (CH_3COO^-) ions. The free hydrogen ions then react with water (H_2O) to form hydronium ions (H_3O^+). This explains the chemical equation above. The longer arrow pointing to the left indicates the tendency for the products—hydronium ions and acetate ions—to re-form acetic acid and water.

Weak acids follow this same pattern of limited ionization in water.

STRONG and WEAK BASES

Strong base – *Example:* Sodium hydroxide (NaOH)

Strong bases ionize completely in water

Other **strong bases**:

• Lithium hydroxide (LiOH)
• Potassium hydroxide (KOH)
• Barium hydroxide (Ba(OH)$_2$)

Weak base – *Example:* Ammonia (NH$_3$)

Weak bases tend *not* to accept hydrogen ions and/or ionize in water; most remain as molecules

Other **weak bases**:

• Trimethylamine [N(CH$_3$)$_3$]
• Pyridine (C$_5$H$_5$N)
• Methylamine (CH$_3$NH$_2$)

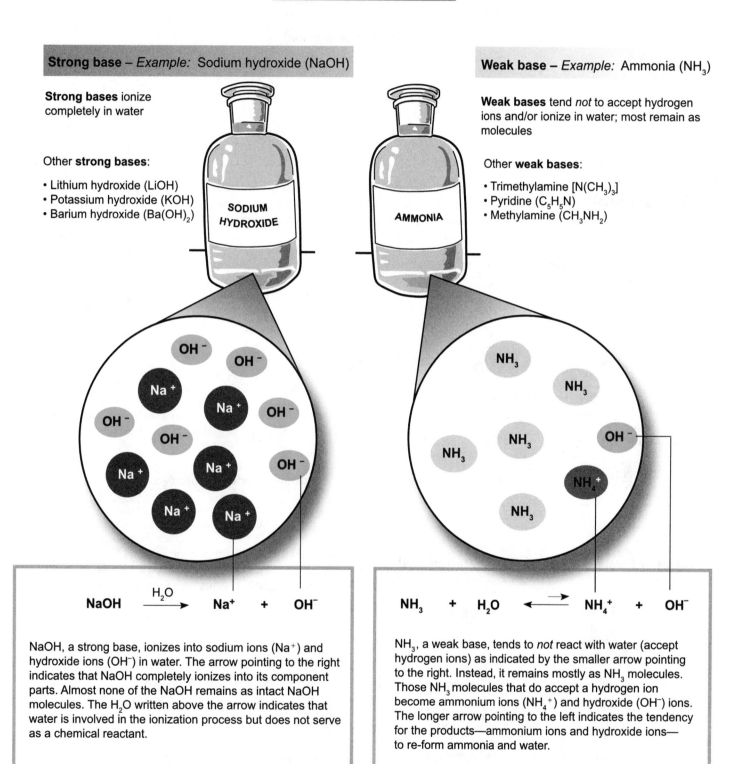

$$NaOH \xrightarrow{H_2O} Na^+ + OH^-$$

$$NH_3 + H_2O \xleftarrow{\longrightarrow} NH_4^+ + OH^-$$

NaOH, a strong base, ionizes into sodium ions (Na$^+$) and hydroxide ions (OH$^-$) in water. The arrow pointing to the right indicates that NaOH completely ionizes into its component parts. Almost none of the NaOH remains as intact NaOH molecules. The H$_2$O written above the arrow indicates that water is involved in the ionization process but does not serve as a chemical reactant.

NH$_3$, a weak base, tends to *not* react with water (accept hydrogen ions) as indicated by the smaller arrow pointing to the right. Instead, it remains mostly as NH$_3$ molecules. Those NH$_3$ molecules that do accept a hydrogen ion become ammonium ions (NH$_4^+$) and hydroxide (OH$^-$) ions. The longer arrow pointing to the left indicates the tendency for the products—ammonium ions and hydroxide ions— to re-form ammonia and water.

Description

Acids and bases undergo a variety of different chemical reactions. Let's examine five common reactions:

1   Acids and metals

2   Acids and metal hydroxides (acid–base reaction)

3   Acids and metal oxides

4   Acids and carbonates or bicarbonates

5   Acids and ammonia or amines

Acids and metals react to produce metal salt and hydrogen gas via a single replacement reaction. Recall that single replacement reactions are like changing dance partners when components recombine to form new products.

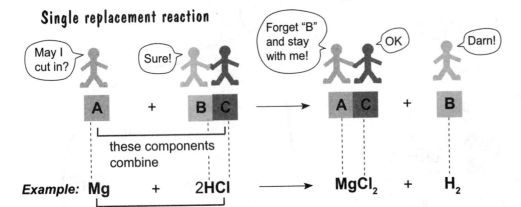

Magnesium ribbon dropped in a test tube containing HCl. Notice the bubbles of H_2 gas.

Magnesium metal reacts with hydrochloric acid to form magnesium chloride and hydrogen gas.

Acids and metal hydroxides (acid–base reaction) react to produce salt and water. Because metal hydroxides are bases, this also is called an acid–base reaction. **Neutralization** occurs—meaning that the resulting solution is neither acidic nor basic.

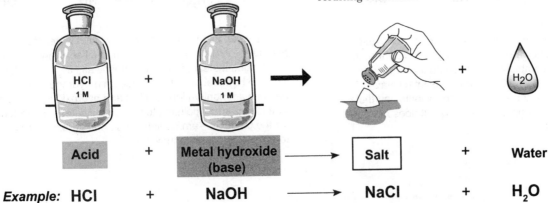

Hydrochloric acid reacts with sodium hydroxide to from sodium chloride and water.

Acids and metal oxides react to produce salt and water.

Sulfuric acid is found in car batteries.

Example: $H_2SO_4 + CuO \longrightarrow CuSO_4 + H_2O$

Sulfuric acid reacts with copper(II) oxide to form copper(II) sulfate and water.

Acids and carbonates or bicarbonates react to produce salt, carbon dioxide, and water.

Carbonates

Example: $2HNO_3 + Na_2CO_3 \longrightarrow 2NaNO_3 + CO_2 + H_2O$

Nitric acid reacts with sodium carbonate to form sodium nitrate and carbon dioxide, and water.

This is the reaction in your stomach when you take an "antacid."

Bicarbonates

Example: $HCl + NaHCO_3 \longrightarrow NaCl + CO_2 + H_2O$

Hydrochloric acid reacts with sodium bicarbonate to form sodium chloride, carbon dioxide, and water.

Acids and ammonia or amines react to produce salt.

Ammonia

Ammonium chloride salt in ionized form

Example: $HCl + NH_3 \longrightarrow NH_4^+ + Cl^-$

Hydrochloric acid reacts with ammonia to form an ammonium ion and a chloride ion.

Amines

Methylammonium chloride salt in ionized form

Example: $HCl + CH_3NH_2 \longrightarrow CH_3NH_3^+ + Cl^-$

Hydrochloric acid reacts with methylamine to form a methylammonium ion and a chloride ion.

Function, Definition, and Examples

Function:

Buffers put a clamp on pH change by preventing large shifts in pH when an acid or a base is added to a buffered solution. Buffers allow a solution to maintain a stable pH at a given pH value.

Definition and Examples:

A buffer consists of a weak acid and its conjugate base. **Buffered solutions** are made by mixing these components with water. **Examples** include:

- Carbonic acid (H_2CO_3) and bicarbonate ion (HCO_3^-)
- Acetic acid (CH_3COOH) and acetate ion (CH_3COO^-)
- Sodium dihydrogen phosphate (NaH_2PO_4) and sodium monohydrogen phosphate (Na_2HPO_4)

Application: Blood pH

— Red blood cell

The normal pH range of blood is a stable **7.35–7.45**. Acids and bases are added to your blood every day from the different foods we ingest and from products of metabolism. Enzymes in the blood that catalyze vital chemical reactions are affected negatively by changes in pH. Death can result if the blood becomes too acidic (< pH 6.8) or too basic (> pH 8.0). Many buffering systems are used to keep the blood pH stable. The most important is the naturally occurring **carbonic acid buffer in our blood**.

How Does a Buffer Work?

Let's examine the **carbonic acid buffering system** in the blood as an example of how a buffer works.

Carbonic acid (H_2CO_3), a weak acid, donates a H^+ to a water molecule to form a hydronium ion (H_3O^+) and a bicarbonate ion (HCO_3^-). The resulting bicarbonate ion is the conjugate base of carbonic acid. A conjugate base is the remaining substance formed whenever a weak acid donates a H^+.

If acid is added to the blood, it increases the H_3O^+ concentration. The bicarbonate ion reacts with the H_3O^+, forming more H_2CO_3, thereby preventing a shift to a more acidic pH:

$$HCO_3^- \ + \ H_3O^+ \longrightarrow H_2CO_3 \ + \ H_2O$$

If a base is added to the blood, it increases the OH^- concentration. The carbonic acid reacts with the OH^-, forming more HCO_3^-, thereby preventing a shift to a more basic pH:

$$H_2CO_3 \ + \ OH^- \longrightarrow HCO_3^- \ + \ H_2O$$

Without a Buffer

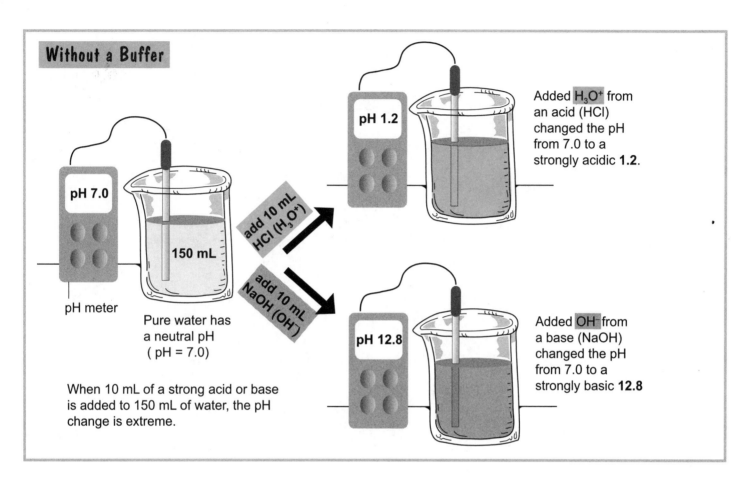

pH 7.0

pH meter

150 mL

Pure water has
a neutral pH
(pH = 7.0)

add 10 mL
HCl (H₃O⁺)

add 10 mL
NaOH (OH⁻)

pH 1.2

Added H₃O⁺ from
an acid (HCl)
changed the pH
from 7.0 to a
strongly acidic **1.2**.

pH 12.8

Added OH⁻ from
a base (NaOH)
changed the pH
from 7.0 to a
strongly basic **12.8**

When 10 mL of a strong acid or base
is added to 150 mL of water, the pH
change is extreme.

With a Buffer

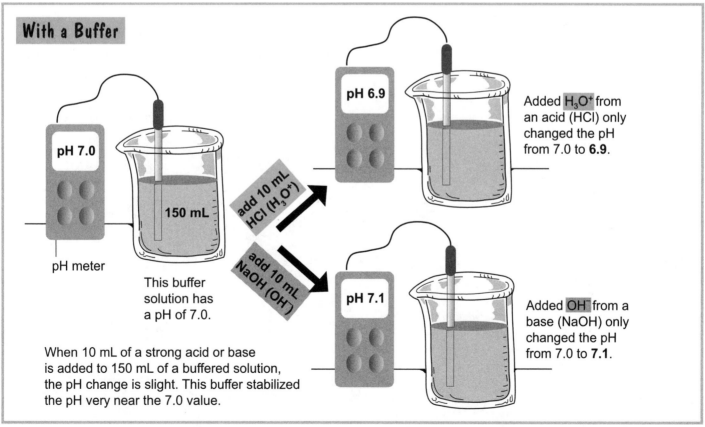

pH 7.0

pH meter

150 mL

This buffer
solution has
a pH of 7.0.

add 10 mL
HCl (H₃O⁺)

add 10 mL
NaOH (OH⁻)

pH 6.9

Added H₃O⁺ from
an acid (HCl) only
changed the pH
from 7.0 to **6.9**.

pH 7.1

Added OH⁻ from a
base (NaOH) only
changed the pH
from 7.0 to **7.1**.

When 10 mL of a strong acid or base
is added to 150 mL of a buffered solution,
the pH change is slight. This buffer stabilized
the pH very near the 7.0 value.

Nuclear
Chemistry

Description

Radiation is energy emitted from an unstable, radioactive nucleus in the form of particles or rays, allowing an unstable nucleus to become more stable. Let's examine six common types of radiation: alpha particle, beta particle, positron, proton, neutron, and gamma rays.

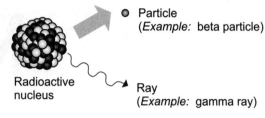

Radioactive nucleus

Particle
(*Example:* beta particle)

Ray
(*Example:* gamma ray)

An **alpha particle (α)** a helium (He) nucleus that contains 2 protons and 2 neutrons; named after the first letter in the Greek alphabet, **alpha (α)**.

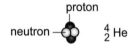

proton

neutron

^4_2He the symbol for an alpha particle representing a helium (He) nucleus with a mass number of 4 and atomic number 2.

Note that the symbol does not show the charge; however, the 2 protons give this particle a $+2$ charge.

A **beta particle (β)** a fast-moving electron; named after the second letter in the Greek alphabet, **beta (β)**.

$^{\ 0}_{-1}e$ the symbol for a beta particle representing an electron with a mass number of 0 and a -1 charge.

A beta particle is formed when a neutron in an unstable nucleus transforms into a proton and an electron.

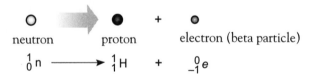

neutron proton + electron (beta particle)

$$^1_0\text{n} \longrightarrow {}^1_1\text{H} + {}^{\ 0}_{-1}e$$

A **positron (β⁺)** a positively charged beta particle, hence β⁺; its name comes from a coined word: positive $+$ -tron (as in electron).

$^{\ 0}_{+1}e$ the symbol for a positron representing a particle with a mass number of 0 and a $+1$ charge.

A positron is essentially a "positive electron."

Proton (p) and neutron (n) — subatomic particles located in the nucleus of an atom.

- ● **Proton**—a subatomic particle

  $_1^1H$ the symbol for a proton representing a hydrogen (H) nucleus with a mass number of 1 and an atomic number of 1.
  Note that the symbol does not show the charge; however, this particle has a $+1$ charge.

- ○ **Neutron**—a subatomic particle

  $_0^1n$ the symbol for a neutron (n) representing a mass number of 1 and a charge of 0. In this case, the symbol does show the charge.

Gamma rays (γ) 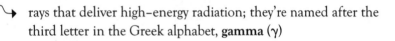 rays that deliver high-energy radiation; they're named after the third letter in the Greek alphabet, **gamma (γ)**

$_0^0\gamma$ the symbol for gamma (γ) rays representing no mass or charge because gamma rays are pure energy.

This table summarizes the six different types of radiation.

	Type of Radiation	Symbol	Symbol meaning	Mass Number	Charge
PARTICLES	Alpha particle (α)	$_2^4He$	**Helium (He) nucleus** with a mass number of 4 and atomic number of 2	4	+2
	Beta particle (β)	$_{-1}^0e$	**Electron (e)** with a mass number of 0 and a charge of -1	0	- 1
	Positron (β⁺)	$_{+1}^0e$	Particle that is **opposite of an electron (e)**; mass number of 0 and a charge of +1	0	+1
	Proton (p)	$_1^1H$	**Hydrogen (H) nucleus** with a mass number of 1 and atomic number of 1	1	+1
	Neutron (n)	$_0^1n$	**Neutron (n)** with a mass number of 1 and a charge of 0	1	0
RAYS	Gamma ray (γ)	$_0^0\gamma$	A ray with no mass or charge	0	0

Description

Radioactive decay is the natural process whereby the nucleus of a radioactive substance emits radiation, resulting in the formation of a different nucleus. In most cases, this new nucleus has a different number of protons and is therefore a new element; however, a more stable form of the original radioactive element also can result. This concept can be written out as an equation:

Concept/equation: Radioactive nucleus ⟶ new nucleus + radiation

Let's examine four different types of radioactive decay: alpha (α) decay, beta (β) decay, positron (β^+) emission, and gamma (γ) emission.

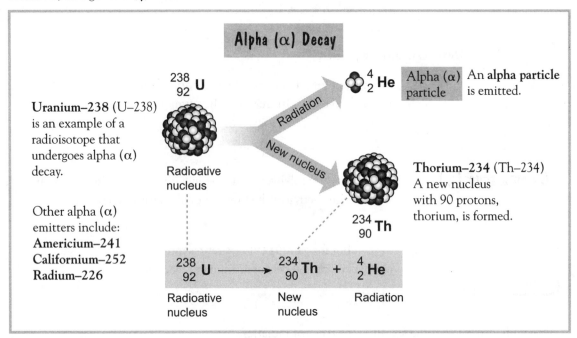

Alpha (α) Decay

Uranium–238 (U–238) is an example of a radioisotope that undergoes alpha (α) decay.

Other alpha (α) emitters include:
Americium–241
Californium–252
Radium–226

$^{238}_{92}$U — Radioative nucleus

Radiation → $^{4}_{2}$He — Alpha (α) particle: An **alpha particle** is emitted.

New nucleus → $^{234}_{90}$Th — **Thorium–234** (Th–234) A new nucleus with 90 protons, thorium, is formed.

$$^{238}_{92}\text{U} \longrightarrow {}^{234}_{90}\text{Th} + {}^{4}_{2}\text{He}$$

Radioative nucleus — New nucleus — Radiation

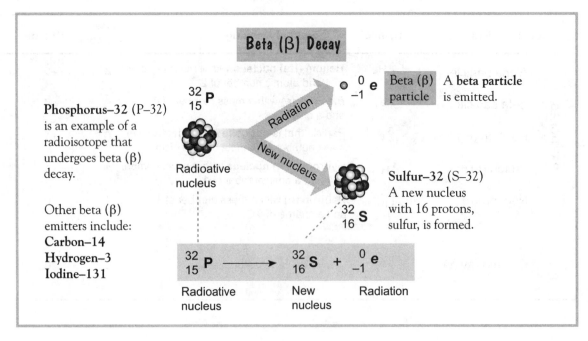

Beta (β) Decay

Phosphorus–32 (P–32) is an example of a radioisotope that undergoes beta (β) decay.

Other beta (β) emitters include:
Carbon–14
Hydrogen–3
Iodine–131

$^{32}_{15}$P — Radioative nucleus

Radiation → $^{0}_{-1}$e — Beta (β) particle: A **beta particle** is emitted.

New nucleus → $^{32}_{16}$S — **Sulfur–32** (S–32) A new nucleus with 16 protons, sulfur, is formed.

$$^{32}_{15}\text{P} \longrightarrow {}^{32}_{16}\text{S} + {}^{0}_{-1}\text{e}$$

Radioative nucleus — New nucleus — Radiation

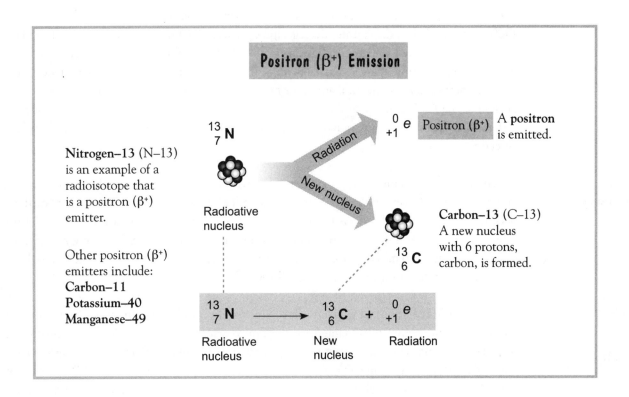

Positron (β⁺) Emission

Nitrogen–13 (N–13) is an example of a radioisotope that is a positron (β⁺) emitter.

Other positron (β⁺) emitters include:
Carbon–11
Potassium–40
Manganese–49

$^{13}_{7}$N

Radioative nucleus

Radiation

New nucleus

$^{0}_{+1}$e   Positron (β⁺)   A **positron** is emitted.

$^{13}_{6}$C

Carbon–13 (C–13) A new nucleus with 6 protons, carbon, is formed.

$$^{13}_{7}\text{N} \longrightarrow {}^{13}_{6}\text{C} + {}^{0}_{+1}\text{e}$$

Radioative nucleus   New nucleus   Radiation

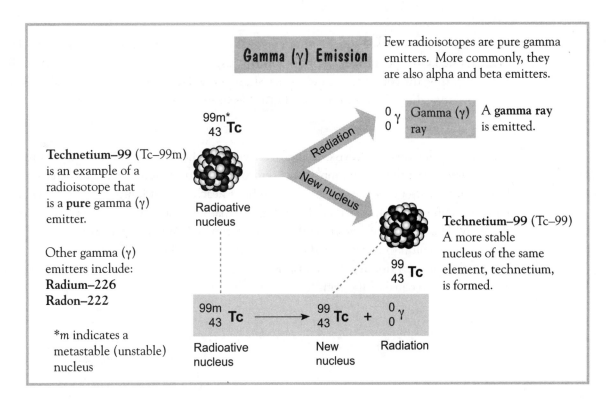

Gamma (γ) Emission

Few radioisotopes are pure gamma emitters. More commonly, they are also alpha and beta emitters.

Technetium–99 (Tc–99m) is an example of a radioisotope that is a **pure** gamma (γ) emitter.

Other gamma (γ) emitters include:
Radium–226
Radon–222

*m indicates a metastable (unstable) nucleus

$^{99m*}_{43}$Tc

Radioative nucleus

Radiation

New nucleus

$^{0}_{0}$γ   Gamma (γ) ray   A **gamma ray** is emitted.

$^{99}_{43}$Tc

Technetium–99 (Tc–99) A more stable nucleus of the same element, technetium, is formed.

$$^{99m}_{43}\text{Tc} \longrightarrow {}^{99}_{43}\text{Tc} + {}^{0}_{0}\gamma$$

Radioative nucleus   New nucleus   Radiation

129

Detecting Radiation

Although it is potentially damaging and dangerous, radiation is not detected by the human senses. Various instruments have been developed to measure radiation. One of the simplest, least expensive, and most portable examples is the Geiger counter, also known as the Geiger–Müller tube. Developed in the early 1900s by Hans Geiger, the counter detects alpha particles, beta particles, and gamma rays. These forms of radiation are called **ionizing radiation**, as they have the ability to separate electrons from atoms or molecules, creating charged particles, ions. In a Geiger counter, ionizing radiation creates ions from an inert gas such as helium, argon, or neon. The resulting charged particles conduct an electrical pulse that then is "counted" as depicted by a meter, lamp, or audible clicks. Geiger counters do not differentiate types of ionizing radiation but, rather, quantify the number of ions produced within the detector.

Units of Measurement

When scientists measure radiation, they use different terms depending on whether they are discussing radiation coming from a radioactive source, the radiation dose absorbed by a person, or the risk for a person's incurring health effects (biological risk) from exposure to radiation. Most scientists in the international community measure radiation using the International System of Units (SI), a uniform system of weights and measures that evolved from the metric system. In the United States, however, the conventional system of measurement is still widely used.

Measuring Emitted Radiation

The amount of radiation being given off, or emitted, by a radioactive material is measured using the conventional unit **curie** (Ci), named for the famed scientist Marie Curie, or the SI unit **becquerel** (Bq). The Ci or Bq is used to express the number of disintegrations of radioactive atoms in a radioactive material over a period of time. For example, one Ci is equal to 37 billion (3.7×10^{10}) disintegrations per second. The Ci is being replaced by the Bq. Because one Bq is equal to one disintegration per second, one Ci is equal to 37 billion (3.7×10^{10}) Bq.

Measuring Radiation Dose

When a person is exposed to radiation, energy is deposited in the tissues of the body. The amount of energy deposited per unit of weight of human tissue is called the absorbed dose. Absorbed dose is measured using the conventional **rad** or the SI unit **gray** (**Gy**). The rad, radiation absorbed dose, was the conventional unit of measurement, but it has been replaced by the **Gy**. One Gy is equal to 100 rad.

Measuring Biological Risk

A person's biological risk (the risk that a person will suffer health effects from an exposure to radiation) is measured using the conventional unit **rem** or the SI unit **sievert** (**Sv**). To determine a person's biological risk, scientists have assigned a number to each type of ionizing radiation (alpha and beta particles, gamma rays, and X-rays) depending on that type's ability to transfer energy to the cells of the body. This number is known as the quality factor (Q). When a person is exposed to radiation, scientists can multiply the dose in rad by the quality factor for the type of radiation present and estimate a person's biological risk in rems. Thus, risk in **rem** = rad × Q. The rem has been replaced by the Sv, and one Sv is equal to 100 rem.

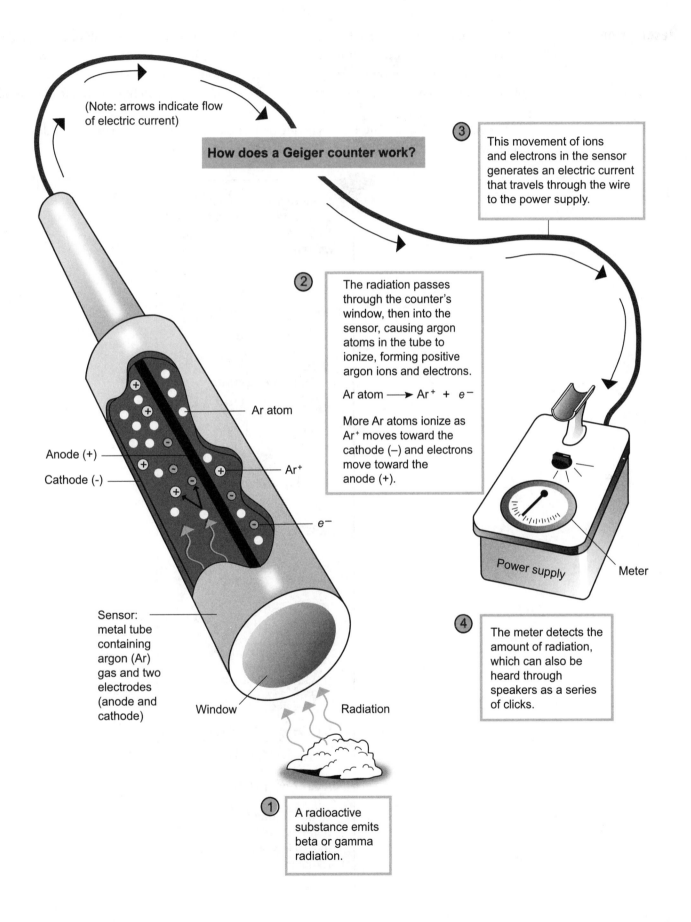

(Note: arrows indicate flow of electric current)

How does a Geiger counter work?

③ This movement of ions and electrons in the sensor generates an electric current that travels through the wire to the power supply.

② The radiation passes through the counter's window, then into the sensor, causing argon atoms in the tube to ionize, forming positive argon ions and electrons.

$$Ar\ atom \longrightarrow Ar^+ + e^-$$

More Ar atoms ionize as Ar$^+$ moves toward the cathode (−) and electrons move toward the anode (+).

Ar atom

Anode (+)

Cathode (-)

Ar$^+$

e^-

Sensor: metal tube containing argon (Ar) gas and two electrodes (anode and cathode)

Window

Radiation

Power supply

Meter

④ The meter detects the amount of radiation, which can also be heard through speakers as a series of clicks.

① A radioactive substance emits beta or gamma radiation.

131

Description

Half–life is the amount of time it takes for a radioactive sample to decay to half of its original value or quantity. It is measured in units of seconds, minutes, hours, days, or years and varies widely depending on the radioisotope. Half–lives can range from a fraction of a second to many years, Polonium–213 has a half–life of 4.2 microseconds, and Uranium–238 has a half–life of 4.46 billion years!

For example, Cesium–136 has a half–life of 13.1 days. This means that after 13.1 days, a 10.0 g sample will decay to half of the original sample, leaving us with 5 g.

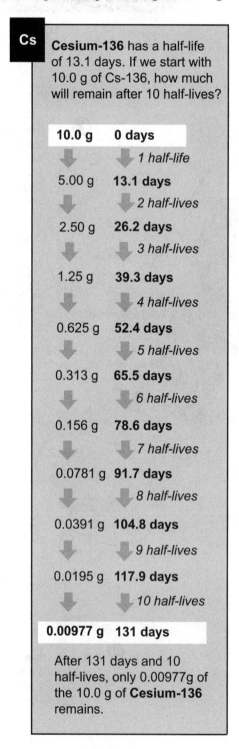

Cesium-136 has a half-life of 13.1 days. If we start with 10.0 g of Cs-136, how much will remain after 10 half-lives?

10.0 g	0 days	
		1 half-life
5.00 g	13.1 days	
		2 half-lives
2.50 g	26.2 days	
		3 half-lives
1.25 g	39.3 days	
		4 half-lives
0.625 g	52.4 days	
		5 half-lives
0.313 g	65.5 days	
		6 half-lives
0.156 g	78.6 days	
		7 half-lives
0.0781 g	91.7 days	
		8 half-lives
0.0391 g	104.8 days	
		9 half-lives
0.0195 g	117.9 days	
		10 half-lives
0.00977 g	**131 days**	

After 131 days and 10 half-lives, only 0.00977g of the 10.0 g of **Cesium-136** remains.

Time frame	Half-lives of different radioisotopes
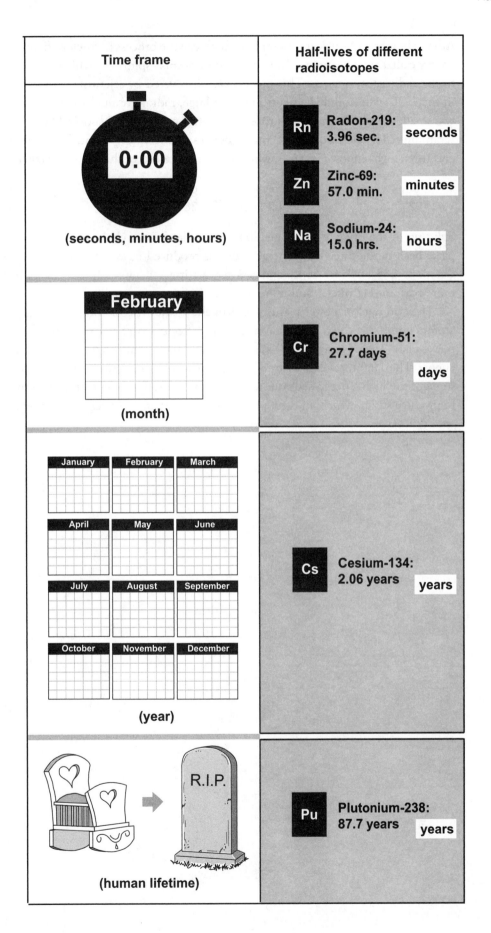(seconds, minutes, hours)	**Rn** Radon-219: 3.96 sec. — seconds **Zn** Zinc-69: 57.0 min. — minutes **Na** Sodium-24: 15.0 hrs. — hours
February (month)	**Cr** Chromium-51: 27.7 days — days
January February March / April May June / July August September / October November December (year)	**Cs** Cesium-134: 2.06 years — years
R.I.P. (human lifetime)	**Pu** Plutonium-238: 87.7 years — years

Description

Atomic bombs and nuclear power plants both use the process of **nuclear fission** to produce a form of energy called **atomic energy**. The term fission means "splitting," which accurately describes changes to the radioactive nucleus. This process was discovered by scientists working in Italy and Germany in the 1930s who collided a neutron with a large, radioative nucleus—Uranium-235—to split it into two smaller nuclei. During this process the neutron actually is absorbed by Uranium-235 to become Uranium-236. The resulting fracture produces one Krypton-91 nucleus, one Barium-142 nucleus, and three high-energy neutrons, which can be expressed in the following equation:

$$^{235}_{92}\text{U} + {}^{1}_{0}\text{n} \longrightarrow {}^{236}_{92}\text{U} \longrightarrow {}^{91}_{36}\text{Kr} + {}^{142}_{56}\text{Ba} + 3{}^{1}_{0}\text{n} + {}^{0}_{0}\gamma + \text{energy}$$

Splitting of the original Uranium-235 nucleus leads to a **chain reaction**, illustrated on the facing page. Each of the three high-energy neutrons produced moves at high speed to collide with another Uranium-235 nucleus. As this process repeats itself, it rapidly produces more Krypton-91 nuclei, more Barium-142 nuclei, and more neutrons in an exponential fashion.

How do nuclear power plants work? In short, they use fission to generate electricity. Unlike an atomic bomb, in which all the atomic energy is released at once as a massive explosion, a nuclear power plant depends on nuclear fission to be slow and controlled to deliver a steady supply of electrical energy. The Uranium-235 is stored in a nuclear reactor, which contains control rods that absorb some of the high-energy neutrons to slow the process. The energy from fission is used to heat water to produce steam, which, in turn, propels a turbine that generates electrical energy.

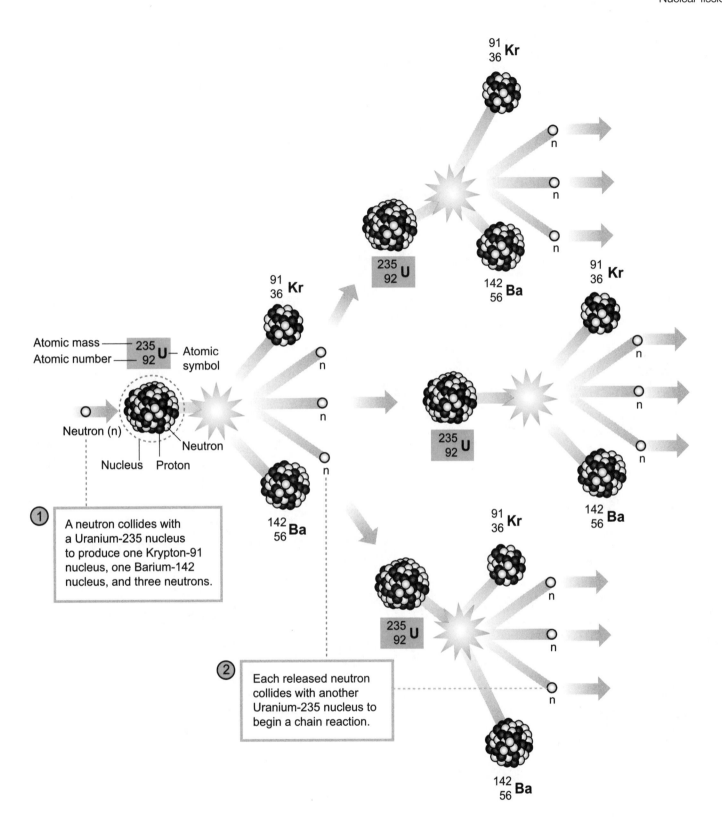

Atomic mass — 235
Atomic number — 92 U — Atomic symbol

Neutron (n)
Nucleus   Proton
Neutron

① A neutron collides with a Uranium-235 nucleus to produce one Krypton-91 nucleus, one Barium-142 nucleus, and three neutrons.

② Each released neutron collides with another Uranium-235 nucleus to begin a chain reaction.

Organic Chemistry

Description

Organic compounds contain the element carbon (C). They typically also contain hydrogen (H) and also may contain other elements such as oxygen (O) and nitrogen (N). Within all organic compounds are groups of atoms called **functional groups** that determine many of the chemical and physical properties of that compound. Functional groups allow us to predict chemical behaviors and thus serve as a basis for classification of the different types of organic compounds. We will examine the following four common types of functional groups: hydroxyl, carbonyl, carboxyl, and amino groups.

Four Common Functional Groups

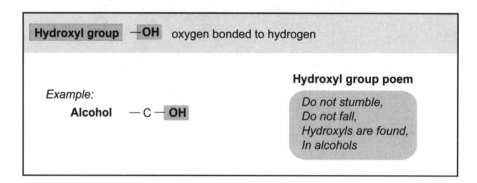

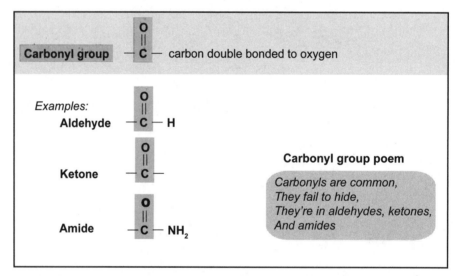

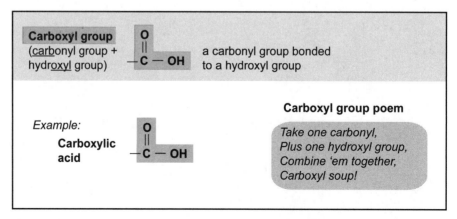

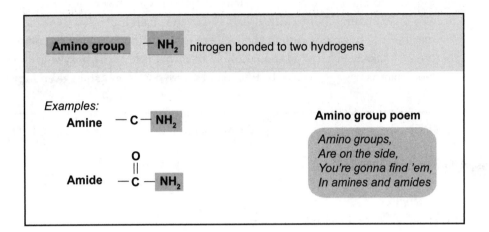

Practice Problems

1 What type of functional group is found in a **ketone**?

2 What type of functional group is found in an **alcohol**?

3 Identify the two different types of functional groups found in an **amide**?

4 What type of functional group is found in a **carboxylic acid**?

Description

Organic compounds contain the element carbon (C). They typically also contain hydrogen (H) and also may contain other elements such as oxygen (O) and nitrogen (N). The bonding of many carbons together results in long carbon chains. Let's examine the following 10 common types of organic compounds: alkane, alkene, alkyne, alcohol, aldehyde, ketone, carboxylic acid, ester, amine, and amide. Note that names are given as **IUPAC** names, with common names given in parentheses.

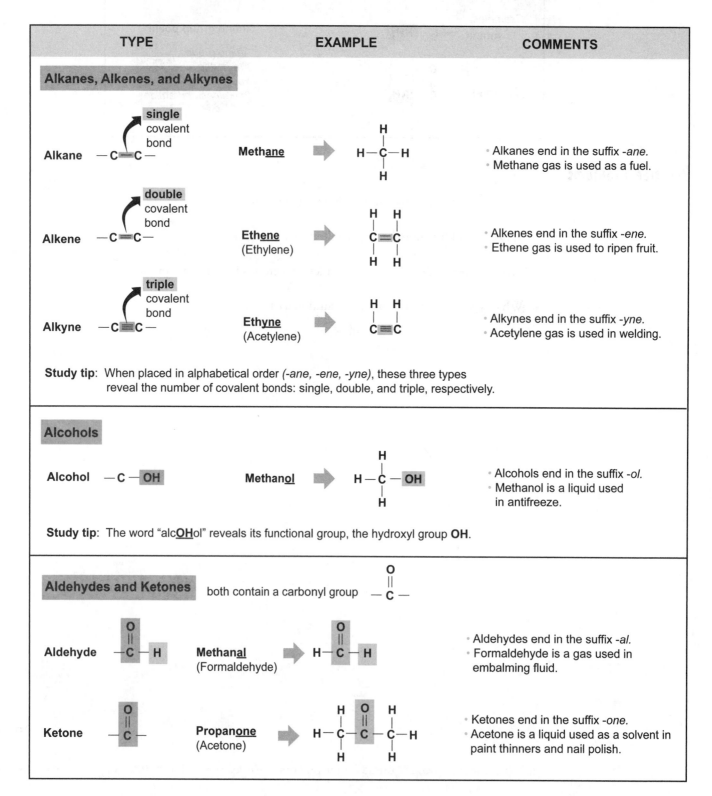

TYPE	EXAMPLE	COMMENTS
Alkanes, Alkenes, and Alkynes		
Alkane	Methane	• Alkanes end in the suffix -ane. • Methane gas is used as a fuel.
Alkene	Ethene (Ethylene)	• Alkenes end in the suffix -ene. • Ethene gas is used to ripen fruit.
Alkyne	Ethyne (Acetylene)	• Alkynes end in the suffix -yne. • Acetylene gas is used in welding.

Study tip: When placed in alphabetical order *(-ane, -ene, -yne)*, these three types reveal the number of covalent bonds: single, double, and triple, respectively.

Alcohols		
Alcohol	Methanol	• Alcohols end in the suffix -ol. • Methanol is a liquid used in antifreeze.

Study tip: The word "alc**OH**ol" reveals its functional group, the hydroxyl group **OH**.

Aldehydes and Ketones both contain a carbonyl group		
Aldehyde	Methanal (Formaldehyde)	• Aldehydes end in the suffix -al. • Formaldehyde is a gas used in embalming fluid.
Ketone	Propanone (Acetone)	• Ketones end in the suffix -one. • Acetone is a liquid used as a solvent in paint thinners and nail polish.

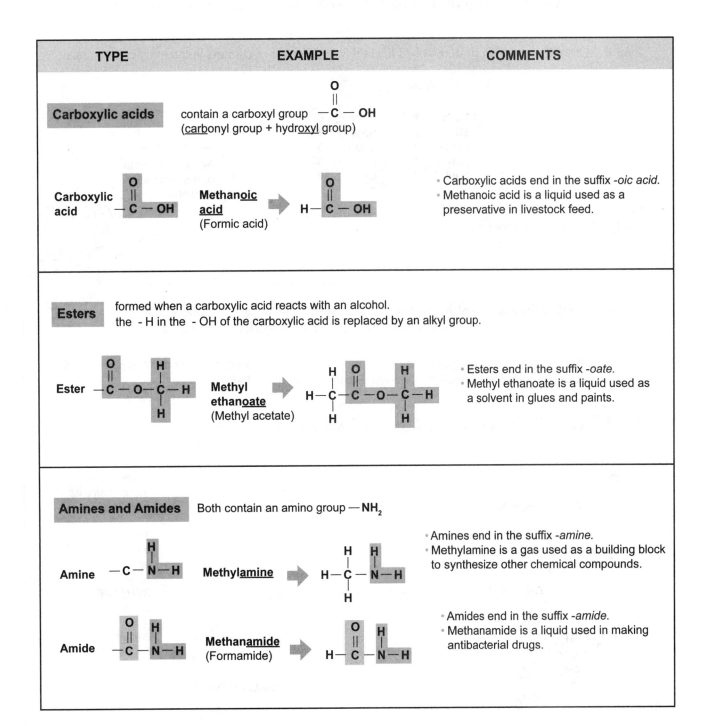

TYPE	EXAMPLE	COMMENTS
Carboxylic acids	contain a carboxyl group $-\overset{\overset{O}{\|\|}}{C}-OH$ (carbonyl group + hydroxyl group)	
Carboxylic acid	$-\overset{\overset{O}{\|\|}}{C}-OH$ **Methanoic acid** (Formic acid) ⟶ $H-\overset{\overset{O}{\|\|}}{C}-OH$	• Carboxylic acids end in the suffix -oic acid. • Methanoic acid is a liquid used as a preservative in livestock feed.
Esters	formed when a carboxylic acid reacts with an alcohol. the - H in the - OH of the carboxylic acid is replaced by an alkyl group.	
Ester	**Methyl ethanoate** (Methyl acetate)	• Esters end in the suffix -oate. • Methyl ethanoate is a liquid used as a solvent in glues and paints.
Amines and Amides	Both contain an amino group $-NH_2$	
Amine	**Methylamine**	• Amines end in the suffix -amine. • Methylamine is a gas used as a building block to synthesize other chemical compounds.
Amide	**Methanamide** (Formamide)	• Amides end in the suffix -amide. • Methanamide is a liquid used in making antibacterial drugs.

Description

Naming organic compounds can be tricky, so it takes some practice. To begin, you need to learn some common prefixes. Then follow the naming system, realizing that it is based on alkane names. You are required to follow a sequence of tasks, so it is best to take a step–by–step approach. First you need to identify the main carbon chain, then you can deal with any branches coming off that chain. When you get organized and understand the system, the naming process becomes easier.

Step 1 Learn the common **prefixes** used to identify the number of carbons in the main carbon chain.

PREFIXES			
Meth-	1	Hex-	6
Eth-	2	Hept-	7
Prop-	3	Oct-	8
But-	4	Non-	9
Pent-	5	Dec-	10

Example:

$CH_3 — CH_3$ ➡ 2 carbons = *eth-*

The **main chain** is like the **main stem** of a plant.

Step 2 | **Identify the main chain** defined as the longest chain of carbons; notice that the main chain does not always follow a linear path.

Note: Three different terms are commonly used to name the main chain: *main chain = longest chain = parent chain.* For consistency, we will use the term "main chain."

$CH_3 — CH_2 — CH_2 — CH_2 — CH_3$ ➡ This **main chain** follows a <u>linear</u> path

CH_3
|
$CH_2 — CH_2 — CH_2 — CH_3$ ➡ This **main chain** has <u>one turn</u>

$CH_3 — CH_2 — CH_2$
|
$CH_2 — CH_3$ ➡ This **main chain** has <u>two turns</u>

$CH_3 — CH_2$
|
$CH_2 — CH_2$
|
CH_3 ➡ This **main chain** has a <u>zig-zag</u> pattern

What do all of these examples have in common?

They are all named **PENTANE** because they contain 5 carbons.

| Step 3 | **Identify all branches coming off the main chain.** |

Note: Three different terms are commonly used for branches coming off the main chain: **branches = side groups = substituent groups.** For consistency, we will use the term "branches."

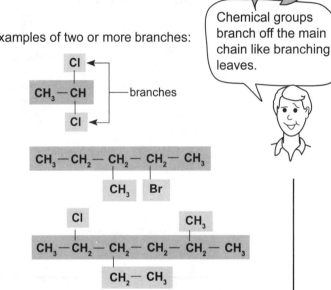

Chemical groups branch off the main chain like branching leaves.

Examples of one branch:

CH$_3$
|
CH$_2$— CH$_2$— CH$_2$— CH$_3$ Cl ← branch

Br
|
CH$_3$— CH$_2$— CH$_3$ ← main chain

CH$_3$— CH$_2$— CH$_2$— CH$_2$— CH$_2$— CH$_3$
|
CH$_3$

Examples of two or more branches:

Cl
|
CH$_3$— CH   branches
|
Cl

CH$_3$— CH$_2$— CH$_2$— CH$_2$— CH$_3$
| |
CH$_3$ Br

Cl   CH$_3$
| |
CH$_3$— CH$_2$— CH$_2$— CH$_2$— CH$_2$— CH$_3$
|
CH$_2$— CH$_3$

| Step 4 | **Identify and name the most common branches.** |

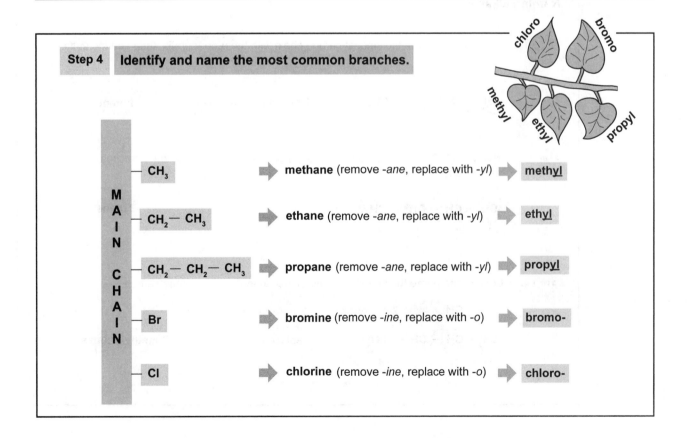

M A I N   C H A I N

— CH$_3$ ⟹ **methane** (remove -*ane*, replace with -*yl*) ⟹ **methyl**

— CH$_2$— CH$_3$ ⟹ **ethane** (remove -*ane*, replace with -*yl*) ⟹ **ethyl**

— CH$_2$— CH$_2$— CH$_3$ ⟹ **propane** (remove -*ane*, replace with -*yl*) ⟹ **propyl**

— Br   ⟹ **bromine** (remove -*ine*, replace with -*o*) ⟹ **bromo-**

— Cl   ⟹ **chlorine** (remove -*ine*, replace with -*o*) ⟹ **chloro-**

Description

Naming organic compounds can be tricky, so it takes some practice. To begin, you need to learn some common prefixes. Then follow the naming based on alkane names. Because you are required to follow a sequence of tasks, it is best to take a step-by-step approach. After you identify the main carbon chain, you can deal with any branches coming off that chain. When you get organized and understand the system, the naming process becomes easier.

First, learn the common prefixes used to identify the number of carbons in the main carbon chain:

PREFIXES			
Meth-	1	Hex-	6
Eth-	2	Hept-	7
Prop-	3	Oct-	8
But-	4	Non-	9
Pent-	5	Dec-	10

Example:

CH_3 ➡ 1 carbon = *meth-*

Second, follow a similar three-step method to name alkanes, alkenes, and alkynes:

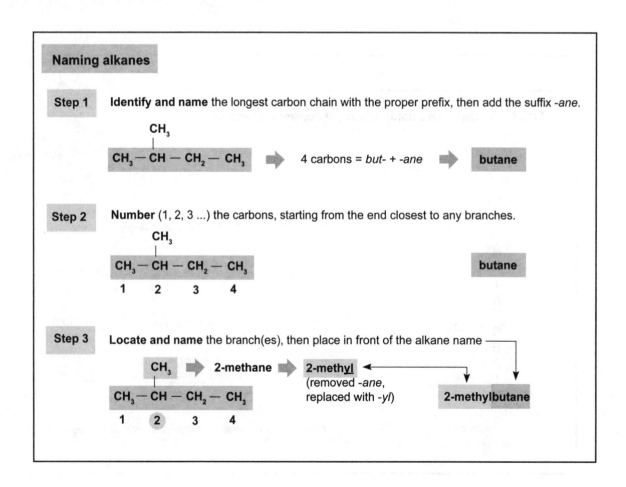

Naming alkanes

Step 1 — **Identify and name** the longest carbon chain with the proper prefix, then add the suffix *-ane*.

$$CH_3 - CH - CH_2 - CH_3 \quad (CH_3)$$ ➡ 4 carbons = *but-* + *-ane* ➡ **butane**

Step 2 — **Number** (1, 2, 3 ...) the carbons, starting from the end closest to any branches.

$$CH_3 - CH - CH_2 - CH_3$$
1   2   3   4

butane

Step 3 — **Locate and name** the branch(es), then place in front of the alkane name

CH_3 ➡ 2-methane ➡ **2-methyl** (removed *-ane*, replaced with *-yl*) ➡ **2-methylbutane**

$$CH_3 - CH - CH_2 - CH_3$$
1   2   3   4

Naming alkenes

Step 1   **Identify and name** the longest carbon chain (that contains the double bond) with the proper prefix, then add the suffix -ene.

5 carbons = *pent-* + *-ene* ➡ **pent<u>ene</u>**

Step 2   **Number** (1, 2, 3 ...) the carbons, starting from the end closest to the double bond; Write the carbon number next to the double bond in front of the alkene name. ─┐

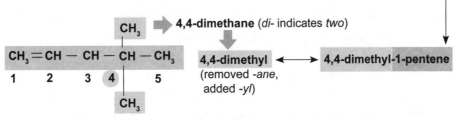

1-pentene

Step 3   **Locate and name** the branch(es), then place in front of the alkene name. ─┐

4,4-dimethane (*di-* indicates *two*)

4,4-dimethyl (removed *-ane*, added *-yl*) ⟷ **4,4-dimethyl-1-pentene**

Naming alkynes

Step 1   **Identify and name** the longest carbon chain (that contains the triple bond) with the proper prefix, then add the suffix -yne.

6 carbons = *hex-* + *-yne* ➡ **hex<u>yne</u>**

Step 2   **Number** (1, 2, 3 ...) the carbons starting from the end closest to the triple bond; Write the carbon number next to the triple bond in front of the alkyne name. ─┐

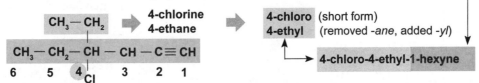

1-hexyne

Step 3   **Locate and name** the branch(es), then place in front of the alkyne name. ─┐

4-chlorine
4-ethane

4-chloro (short form)
4-ethyl (removed *-ane*, added *-yl*)

➡ **4-chloro-4-ethyl-1-hexyne**

Description

Naming organic compounds can be tricky, and it takes some practice. To begin, you have to learn some common prefixes. Then follow the naming based on alkane names. You are required to follow a sequence of tasks, so it is best to take a step-by-step approach. First, identify the main carbon chain, then you can deal with any branches coming off that chain. When you get organized and understand the system, the naming process becomes easier.

First, learn the common prefixes used to identify the number of carbons in the main carbon chain:

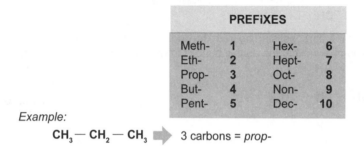

PREFIXES			
Meth-	1	Hex-	6
Eth-	2	Hept-	7
Prop-	3	Oct-	8
But-	4	Non-	9
Pent-	5	Dec-	10

Example:

$CH_3 — CH_2 — CH_3$ ➡ 3 carbons = *prop-*

Second, follow a similar three-step method to name alcohols, aldehydes, and ketones:

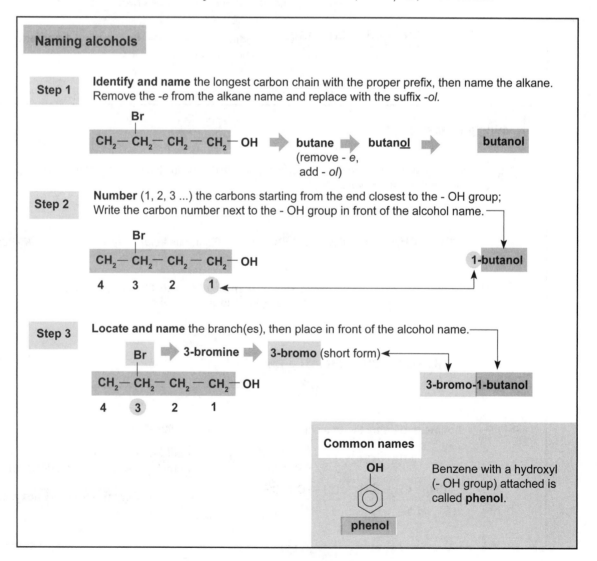

Naming alcohols

Step 1 **Identify and name** the longest carbon chain with the proper prefix, then name the alkane. Remove the -e from the alkane name and replace with the suffix -ol.

$CH_2 — CH_2 — CH_2 — CH_2 — OH$ ➡ butane ➡ butanol (remove - *e*, add - *ol*) ➡ butanol

Step 2 **Number** (1, 2, 3 ...) the carbons starting from the end closest to the - OH group; Write the carbon number next to the - OH group in front of the alcohol name.

1-butanol

Step 3 **Locate and name** the branch(es), then place in front of the alcohol name.

Br ➡ 3-bromine ➡ 3-bromo (short form)

3-bromo-1-butanol

Common names

Benzene with a hydroxyl (- OH group) attached is called **phenol**.

phenol

146

Naming aldehydes

Step 1 **Identify and name** the longest carbon chain (that contains the carbonyl group) with the proper prefix, then name the alkane. Remove the -e and replace it with the suffix -al. No number is used with the main chain.

pentane (remove -e, add -al) → pentan**al** → pentan**al**

Step 2 **Locate and name** the branch(es) by counting the carbonyl carbon as carbon #1.

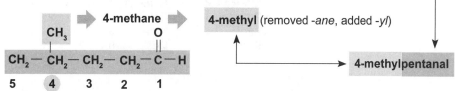

4-methane → **4-methyl** (removed -ane, added -yl)

4-methyl**pentanal**

Common names

The first four simplest, unbranched aldehydes have common names ending with the suffix -aldehyde. The prefix describes the carboxylic acid from which each was derived.

Common names

methanal	=	form*aldehyde*
ethanal	=	acet*aldehyde*
propanal	=	proprion*aldehyde*
butanal	=	butyr*aldehyde*

Naming ketones

Step 1 **Identify and name** the longest carbon chain (that contains the carbonyl group) with the proper prefix, then name the alkane. Remove the -e and replace it with the suffix -one.

butane → butan**one** → butan**one**

(remove -e, add -one)

Step 2 **Number** (1, 2, 3 ...) the carbons starting from the end closest to the carbonyl group; Write the carbonyl carbon number in front of the ketone name.

2-butanone

Step 3 **Locate and name** the branch(es), then place in front of the ketone name.

4-chlorine → **4-chloro** (short form)

4-chloro-2-butanone

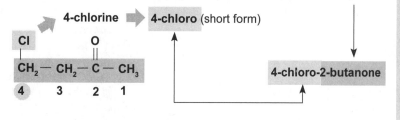

Cyclic ketones

Add the prefix cyclo- in front of the ketone name

cyclohexanone

Description

Naming organic compounds can be tricky and takes some practice. To begin, you need to learn some common prefixes. Then follow the naming based on alkane names. You are required to follow a sequence of tasks, so it is best to take a step-by-step approach. First identify the main carbon chain, then you can deal with any branches coming off that chain. When you get organized and understand the system, the naming process becomes easier.

First, learn the common prefixes used to identify the number of carbons in the main carbon chain:

PREFIXES			
Meth-	1	Hex-	6
Eth-	2	Hept-	7
Prop-	3	Oct-	8
But-	4	Non-	9
Pent-	5	Dec-	10

Example:

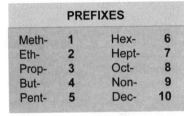

$CH_3 — CH_2 — CH_2 — CH_3$ ➡ 4 carbons = *but-*

Second, follow a similar three-step method to name carboxylic acids and esters:

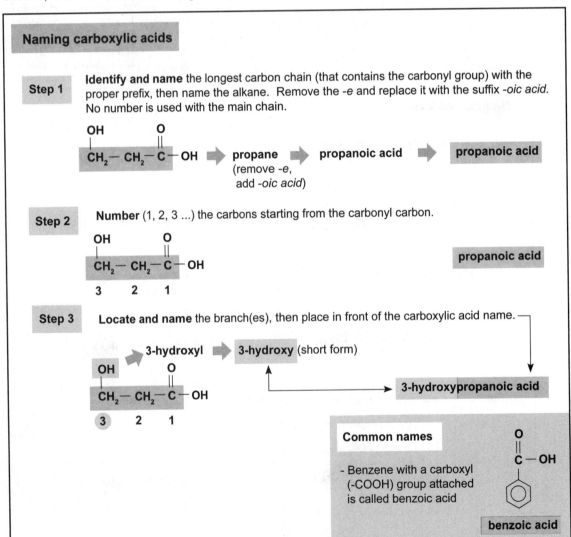

Naming carboxylic acids

Step 1   **Identify and name** the longest carbon chain (that contains the carbonyl group) with the proper prefix, then name the alkane. Remove the *-e* and replace it with the suffix *-oic acid*. No number is used with the main chain.

$CH_2 — CH_2 — C — OH$ ➡ propane ➡ propanoic acid ➡ **propanoic acid**
(remove *-e*, add *-oic acid*)

Step 2   **Number** (1, 2, 3 ...) the carbons starting from the carbonyl carbon.

$CH_2 — CH_2 — C — OH$
3   2   1

propanoic acid

Step 3   **Locate and name** the branch(es), then place in front of the carboxylic acid name.

3-hydroxyl ➡ **3-hydroxy** (short form)

$CH_2 — CH_2 — C — OH$
3   2   1

3-hydroxypropanoic acid

Common names

- Benzene with a carboxyl (-COOH) group attached is called benzoic acid

$C — OH$

benzoic acid

Naming esters

Esters have a two-part name. Recall that esters are formed from reacting an **alcohol** with a **carboxylic acid**.
The first part of the name is based on the alcohol name, and the second part is based on the carboxylic acid name.

$$CH_3 - CH_2 - CH_2 - OH \quad + \quad HO - \overset{\overset{\displaystyle O}{\|}}{C} - CH_3 \quad \longrightarrow$$

alcohol
(1-propanol)

carboxylic acid
(ethanoic acid)

$$CH_3 - CH_2 - CH_2 - O - \overset{\overset{\displaystyle O}{\|}}{C} - CH_3 \quad + \quad H_2O$$

ester

Step 1   **Identify and name** the **alcohol** from which the ester was derived.

$$CH_3 - CH_2 - CH_2 - O - \overset{\overset{\displaystyle O}{\|}}{C} - CH_3 \quad \Rightarrow$$

propanol
(remove -*anol*,
add -*yl*)

$\Rightarrow$ **propyl**

Step 2   **Identify and name** the **carboxylic acid** from which the ester was derived.

$$CH_3 - CH_2 - CH_2 - O - \overset{\overset{\displaystyle O}{\|}}{C} - CH_3 \quad \Rightarrow$$

ethanoic acid
(remove -*oic acid*,
add -*oate*)

$\Rightarrow$ **ethanoate**

Step 3   Put both names together. List the alcohol derivative first, followed by the carboxylic acid.

$$CH_3 - CH_2 - CH_2 - O - \overset{\overset{\displaystyle O}{\|}}{C} - CH_3 \quad \Rightarrow$$

propyl   **ethanoate**

Common names

The first part of the name remains the same.
The second part changes and ends in the
suffix -*ate*.

Example:
propyl ethanoate = propyl acet<u>ate</u>

Common names
for the second part
of the ester name

methanoate	= form*ate*
ethanoate	= acet*ate*
propanoate	= propion*ate*
butanoate	= butyr*ate*

Description

Naming organic compounds can be tricky, so it takes some practice. To begin, you need to learn some common prefixes. Then follow the naming based on alkane names. Because you are required to follow a sequence of tasks, it is best to take a step-by-step approach. First identify the main carbon chain, then you can deal with any branches coming off that chain. When you get organized and understand the system, the naming process becomes easier.

First, learn the common prefixes used to identify the number of carbons in the main carbon chain:

PREFIXES			
Meth-	1	Hex-	6
Eth-	2	Hept-	7
Prop-	3	Oct-	8
But-	4	Non-	9
Pent-	5	Dec-	10

Example:

$CH_3 - CH_3$ ➡ 2 carbons = *eth-*

Second, follow the step-by-step method to name amines and amides:

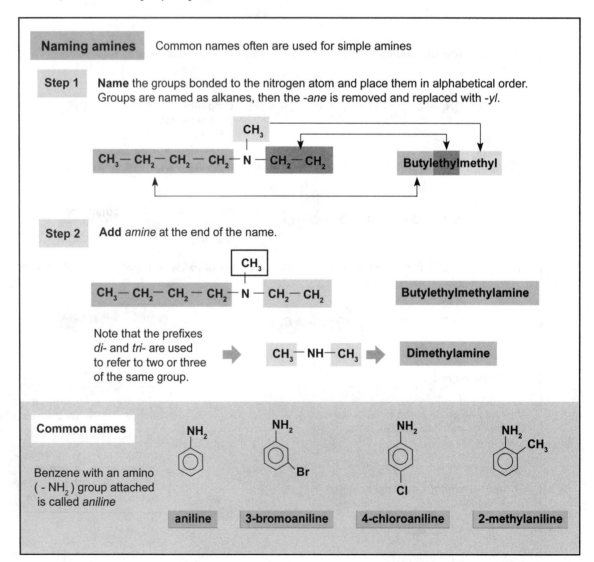

Naming amines Common names often are used for simple amines

Step 1 **Name** the groups bonded to the nitrogen atom and place them in alphabetical order. Groups are named as alkanes, then the *-ane* is removed and replaced with *-yl*.

CH_3

$CH_3 - CH_2 - CH_2 - CH_2 - N - CH_2 - CH_2$ **Butylethylmethyl**

Step 2 **Add** *amine* at the end of the name.

CH_3

$CH_3 - CH_2 - CH_2 - CH_2 - N - CH_2 - CH_2$ **Butylethylmethylamine**

Note that the prefixes *di-* and *tri-* are used to refer to two or three of the same group. ➡ $CH_3 - NH - CH_3$ ➡ **Dimethylamine**

Common names

Benzene with an amino (-NH_2) group attached is called *aniline*

aniline **3-bromoaniline** **4-chloroaniline** **2-methylaniline**

Naming amides

The naming is based on the carboxylic acid name from which the amide was derived.

Step 1

Identify and name the carboxylic acid from which the amide was derived.

$$CH_3-\overset{\overset{\displaystyle O}{\|}}{C}-OH$$

carboxylic acid
(ethanoic acid)

↓

$$CH_3-\overset{\overset{\displaystyle O}{\|}}{C}-NH_2$$

amide

Step 2

Remove the *-oic acid* from the carboxylic acid name and **replace** with the suffix *-amide*.

$$CH_3-\overset{\overset{\displaystyle O}{\|}}{C}-NH_2$$ ➡ **ethanamide**

Examples of other amides

$$H-\overset{\overset{\displaystyle O}{\|}}{C}-NH_2$$

methanamide

$$CH_3-CH_2-\overset{\overset{\displaystyle O}{\|}}{C}-NH_2$$

propanamide

$$\overset{\overset{\displaystyle O}{\|}}{C}-NH_2$$ (benzene ring)

benzamide

- *benz*ene ring
+ *-amide*

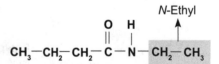

N-Ethyl

N-ethylbutanamide

- groups attached to
the nitrogen of an amide
are named using the
prefix *N*-

Common names

The first four, simplest amides
have common names ending
with the suffix *-amide*. The prefix
describes the carboxylic acid from
which each was derived.

Common names

methanamide	=	form*amide*
ethanamide	=	acet*amide*
propanamide	=	proprion*amide*
butanamide	=	butyr*amide*

Description

Polymers are large, long-chain molecules formed by linking building blocks called **monomers**. Think of a children's necklace that is formed by pushing plastic beads together. In this analogy, the beads are the monomers and the necklace is the polymer. Some polymers are made from the same monomer repeated again and again. This is indicated here with beads having the same color. Other polymers are made from more than one type of monomer, as indicated with beads of different colors. Synthetic polymers, derived from alkenes, are the foundation of the plastics industry. In contrast, our bodies have the ability to make natural polymers, and they also are found in many of the foods we ingest. Let's examine both synthetic and natural polymers.

Analogy

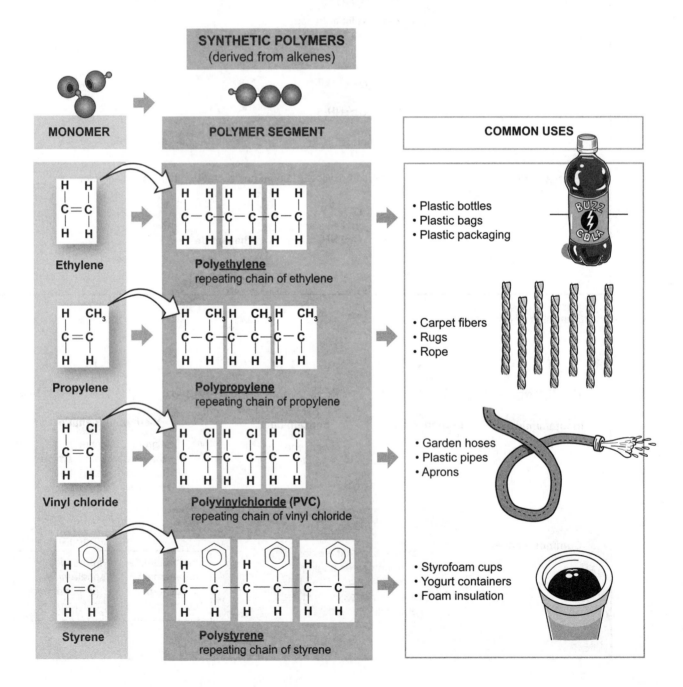

Natural Polymers

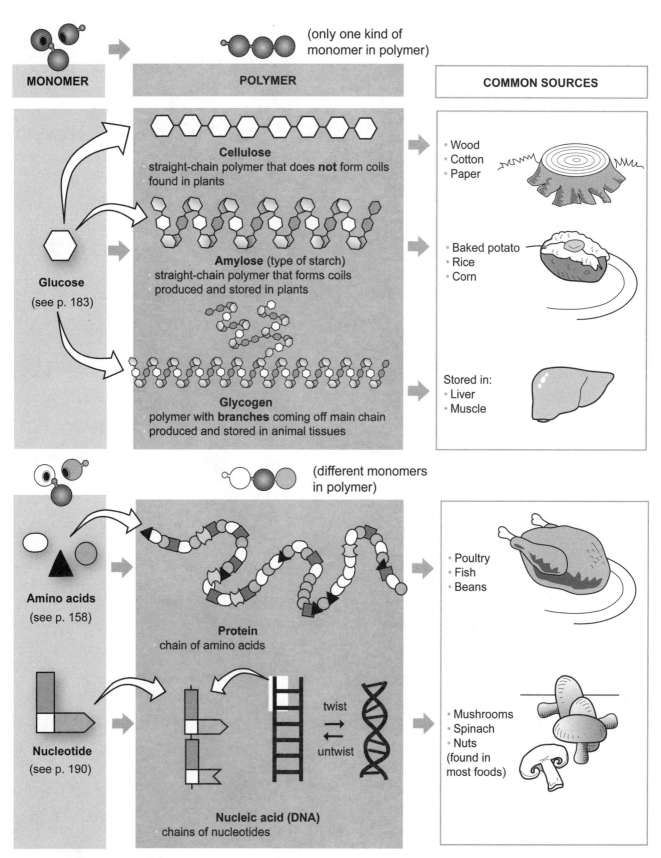

(only one kind of monomer in polymer)

MONOMER

POLYMER

COMMON SOURCES

Cellulose
straight-chain polymer that does **not** form coils found in plants

Glucose
(see p. 183)

Amylose (type of starch)
straight-chain polymer that forms coils produced and stored in plants

Glycogen
polymer with **branches** coming off main chain produced and stored in animal tissues

• Wood
• Cotton
• Paper

• Baked potato
• Rice
• Corn

Stored in:
• Liver
• Muscle

(different monomers in polymer)

Amino acids
(see p. 158)

Protein
chain of amino acids

Nucleotide
(see p. 190)

twist ⇌ untwist

Nucleic acid (DNA)
chains of nucleotides

• Poultry
• Fish
• Beans

• Mushrooms
• Spinach
• Nuts
(found in most foods)

Amino Acids, Polypeptides, and Proteins

Description

Let's examine the chemical structure of **amino acids**—the building blocks of **proteins**.

Recall that **polymers** are large, long-chain molecules formed by linking building blocks called **monomers**. Proteins are natural polymers formed from amino acid monomers. It's like how a children's necklace is formed by pushing plastic beads together. In this analogy, the beads are the amino acid monomers and the necklace is the protein polymer.

Analogy

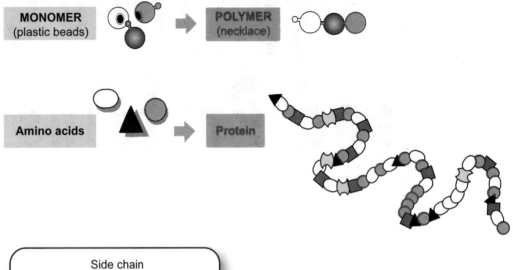

General Structure

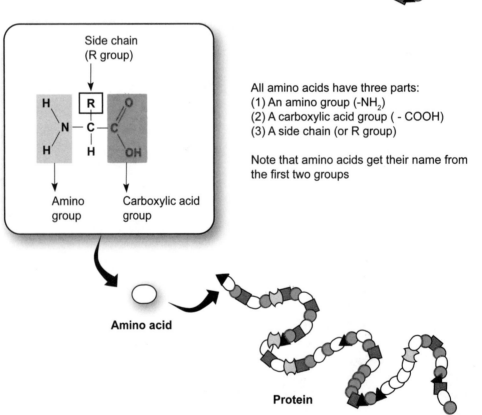

All amino acids have three parts:
(1) An amino group ($-NH_2$)
(2) A carboxylic acid group (- COOH)
(3) A side chain (or R group)

Note that amino acids get their name from the first two groups

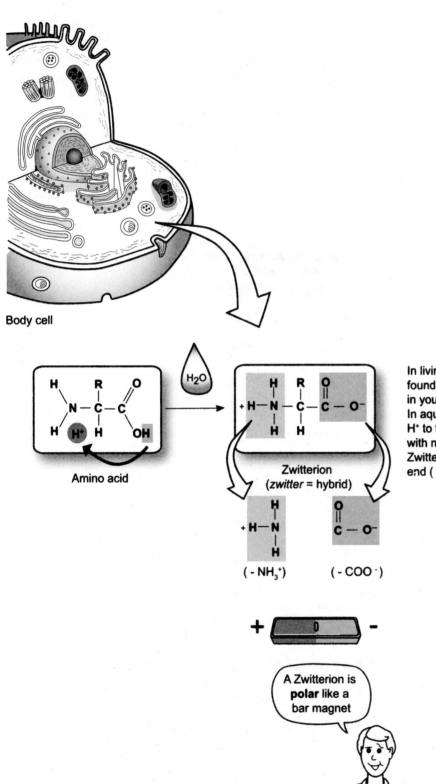

Body cell

Amino acid

Zwitterion
(*zwitter* = hybrid)

$(-NH_3^+)$

$(-COO^-)$

In living systems, amino acids typically are found in aqueous solutions such as the plasma in your blood or the fluid in your body cells. In aqueous solutions, the - COOH donates an H^+ to the - NH_2 to form a Zwitterion—an ion with no net charge. Like a bar magnet, the Zwitterion is polar because it has a positive (+) end (- NH_3^+) and a negative (-) end (- COO^-).

A Zwitterion is **polar** like a bar magnet

Description

The 20 different amino acids are divided into four categories based on their chemical properties: non-polar (9), polar (6), basic (3), and acidic (2). To understand how one amino acid differs from another, recall its general structure. Note that the structure depicted on the facing page is the predominant form at a pH of 7.4. This is the same pH as blood plasma, which is significant because amino acids are transported through the blood. The R group, or side chain, is the portion that changes while the rest of the structure remains the same. Note that the R group has been highlighted in color to allow for easy comparisons.

List

Here is a list of the 20 different kinds of amino acids, in alphabetical order, with their three-letter abbreviations.

The 20 Amino Acids and their Abbreviations (alphabetical order)	
Alanine	Ala
Arginine	Arg
Asparagine	Asn
Aspartic acid	Asp
Cysteine	Cys
Glutamic acid	Glu
Glutamine	Gln
Glycine	Gly
Histidine	His
Isoleucine	Ile
Leucine	Leu
Lysine	Lys
Methionine	Met
Phenylalanine	Phe
Proline	Pro
Serine	Ser
Threonine	Thr
Tryptophan	Trp
Tyrosine	Tyr
Valine	Val

Nonpolar amino acids have nonpolar side chains such as alkyl groups or aromatic groups; hydrophobic ("water fearing")

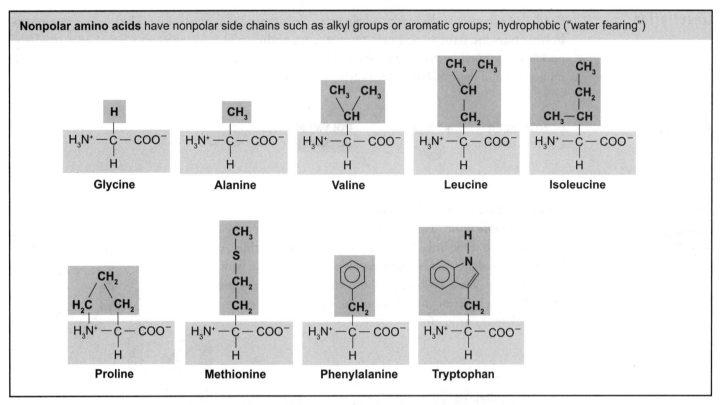

Glycine   Alanine   Valine   Leucine   Isoleucine

Proline   Methionine   Phenylalanine   Tryptophan

Polar amino acids contain polar side chains such as hydroxyl (-OH) groups; hydrophilic ("water loving")

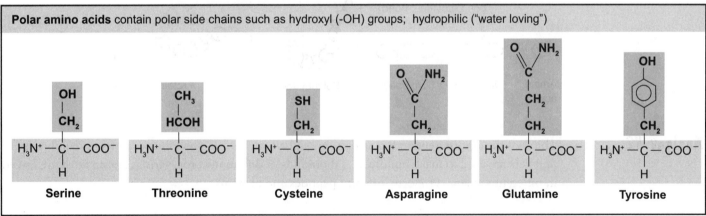

Serine   Threonine   Cysteine   Asparagine   Glutamine   Tyrosine

Basic amino acids contain amine (- NH₂) groups

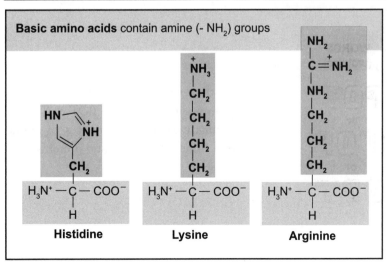

Histidine   Lysine   Arginine

Acidic amino acids contain carboxylic acid (-COOH) groups

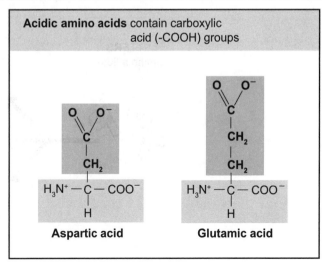

Aspartic acid   Glutamic acid

159

AMINO ACIDS, POLYPEPTIDES, AND PROTEINS

Formation of a polypeptide or a protein

Description

Recall that **polymers** are large, long–chain molecules formed by linking building blocks called **monomers**. **Polypeptides** are natural polymers formed from **amino acid** monomers. It's like how a children's necklace is formed by pushing plastic beads together. In this analogy, the beads are the amino acid monomers and the necklace is the polypeptide polymer.

Analogy #1

Overview: Building the Chain

When one amino acid links to another it forms a C–N covalent bond called a **peptide bond**. Two amino acids linked together are called a **dipeptide**, and three linked together are called a **tripeptide**. As the chain grows in length, it becomes a **polypeptide**. A chain with 50 or more amino acids *and* biological function is called a **protein**. (*Note:* There is no clear consensus on the exact number of amino acids to define a protein).

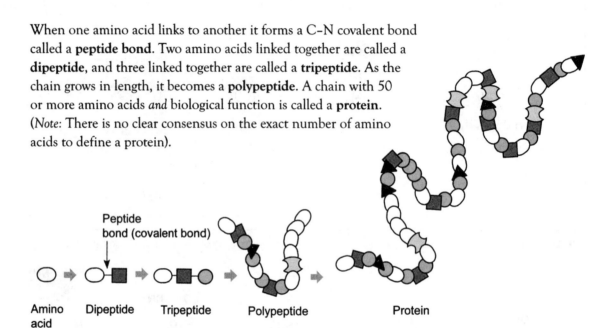

Analogy #2

Think of the 20 different amino acids as an alphabet with 20 letters. Just as different combinations of letters are used to make hundreds of thousands of different words, various amino acids are linked in different sequences to form different peptides and proteins.

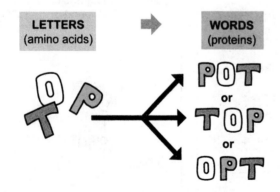

The same three letters—O, T, and P—can be used to form three different words: POT or TOP or OPT. Similarly, different sequences of amino acids form different peptides and proteins.

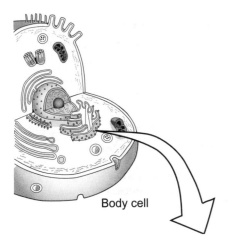

Body cell

Forming the Peptide Bond

The process of creating proteins from amino acids occurs in body cells and is called **protein synthesis** (see p. 200). As a preview, let's look at the formation of a peptide bond between two amino acids. Using the same icons on the facing page, imagine we are forming a dipeptide. When one amino acid links to another, it forms a dipeptide and one water molecule is produced as a product:

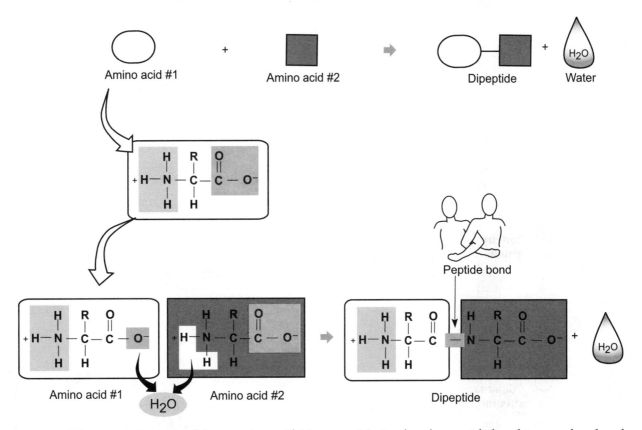

The oxygen is removed from amino acid #1, which combines with two hydrogens from amino acid #2 to form H_2O. The carbon and nitrogen form a peptide bond.

Notice that the peptide bond is a covalent bond between the carbon (C) of amino acid #1 and the nitrogen (N) of amino acid #2. Because all covalent bonds are a shared pair of electrons, it is like two people locked arm in arm.

To form a tripeptide, and then a polypeptide, the process simply repeats itself.

Description

Proteins often have highly complex structures. From simplest to most complex, the four levels of protein structure are: primary structure, secondary structure, tertiary structure, and quaternary structure.

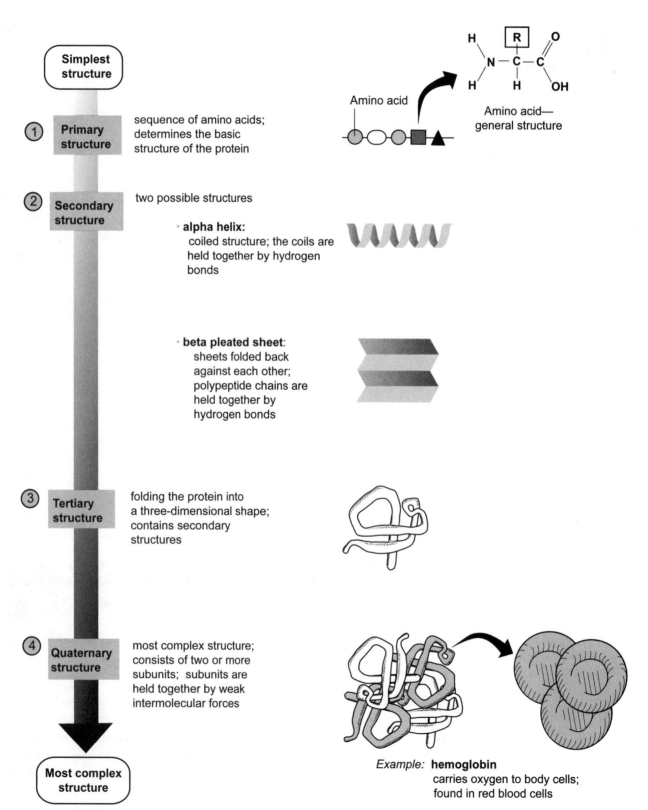

Simplest structure

① **Primary structure** — sequence of amino acids; determines the basic structure of the protein

Amino acid

Amino acid— general structure

② **Secondary structure** — two possible structures

- **alpha helix:** coiled structure; the coils are held together by hydrogen bonds

- **beta pleated sheet:** sheets folded back against each other; polypeptide chains are held together by hydrogen bonds

③ **Tertiary structure** — folding the protein into a three-dimensional shape; contains secondary structures

④ **Quaternary structure** — most complex structure; consists of two or more subunits; subunits are held together by weak intermolecular forces

Most complex structure

Example: **hemoglobin** carries oxygen to body cells; found in red blood cells

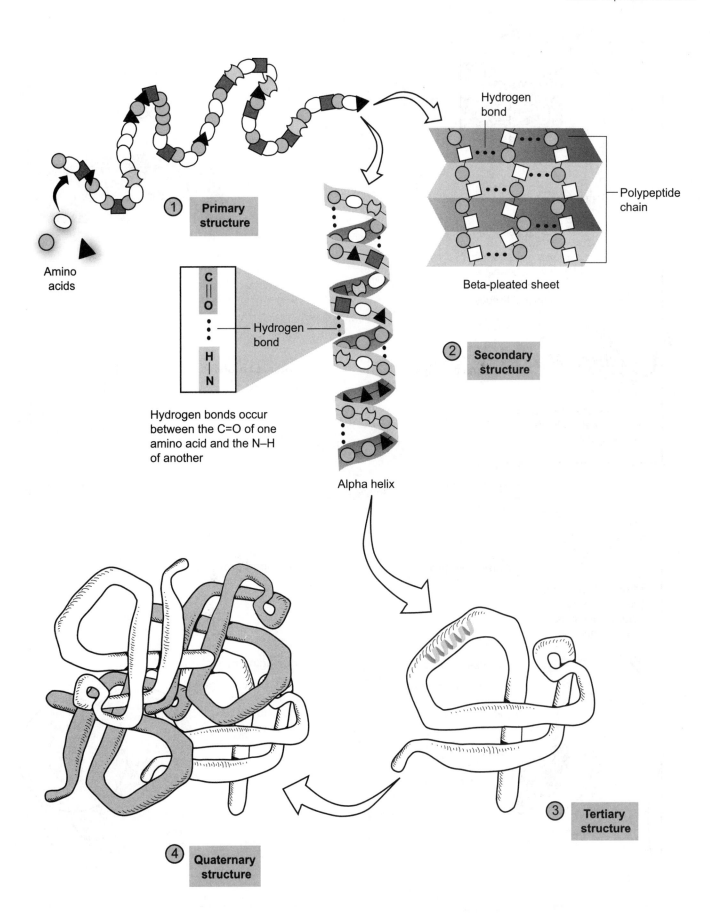

Amino acids

① **Primary structure**

C=O ⋯ H–N

Hydrogen bond

Hydrogen bonds occur between the C=O of one amino acid and the N–H of another

Alpha helix

Hydrogen bond

Polypeptide chain

Beta-pleated sheet

② **Secondary structure**

③ **Tertiary structure**

④ **Quaternary structure**

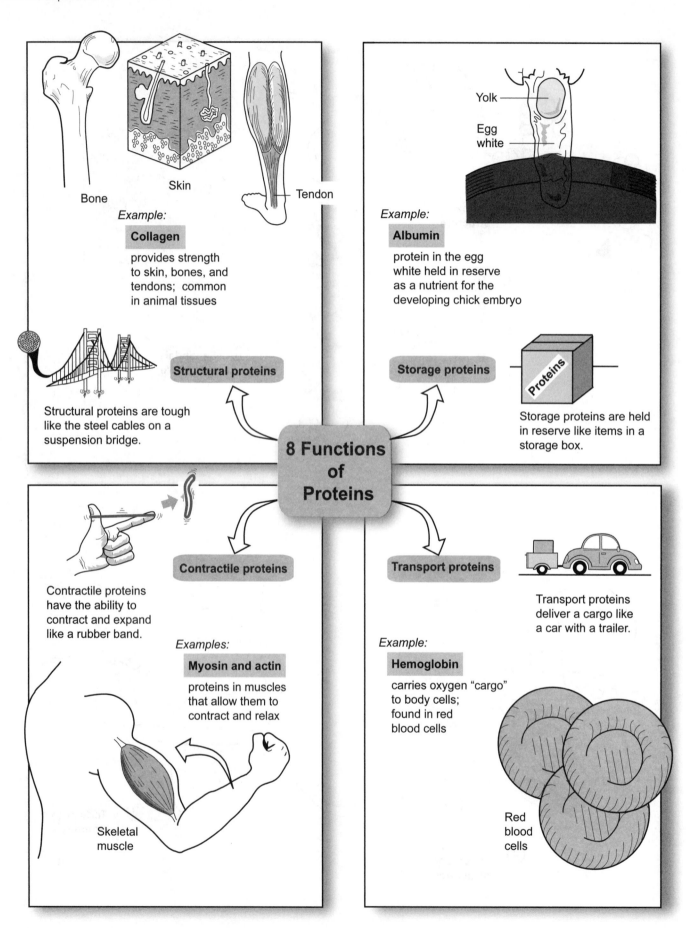

Bone

Skin

Tendon

Example:

Collagen

provides strength to skin, bones, and tendons; common in animal tissues

Structural proteins

Structural proteins are tough like the steel cables on a suspension bridge.

Example:

Yolk

Egg white

Albumin

protein in the egg white held in reserve as a nutrient for the developing chick embryo

Storage proteins

Proteins

Storage proteins are held in reserve like items in a storage box.

8 Functions of Proteins

Contractile proteins

Contractile proteins have the ability to contract and expand like a rubber band.

Examples:

Myosin and actin

proteins in muscles that allow them to contract and relax

Skeletal muscle

Transport proteins

Transport proteins deliver a cargo like a car with a trailer.

Example:

Hemoglobin

carries oxygen "cargo" to body cells; found in red blood cells

Red blood cells

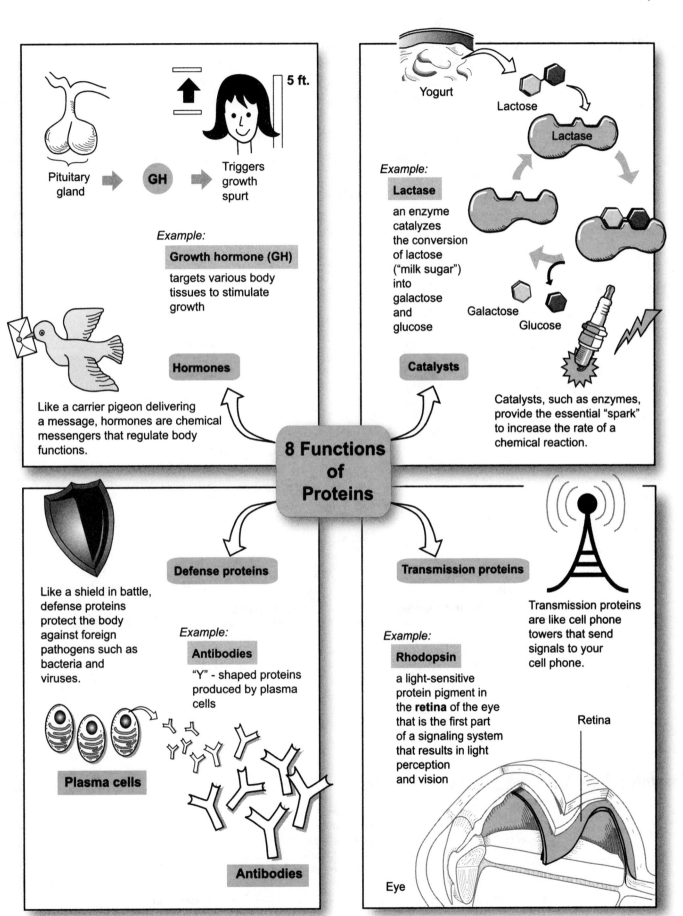

5 ft.

Pituitary gland → **GH** → Triggers growth spurt

Example:

Growth hormone (GH)

targets various body tissues to stimulate growth

Hormones

Like a carrier pigeon delivering a message, hormones are chemical messengers that regulate body functions.

Yogurt

Lactose

Lactase

Example:

Lactase

an enzyme catalyzes the conversion of lactose ("milk sugar") into galactose and glucose

Galactose

Glucose

Catalysts

Catalysts, such as enzymes, provide the essential "spark" to increase the rate of a chemical reaction.

8 Functions of Proteins

Defense proteins

Like a shield in battle, defense proteins protect the body against foreign pathogens such as bacteria and viruses.

Example:

Antibodies

"Y" - shaped proteins produced by plasma cells

Plasma cells

Antibodies

Transmission proteins

Transmission proteins are like cell phone towers that send signals to your cell phone.

Example:

Rhodopsin

a light-sensitive protein pigment in the **retina** of the eye that is the first part of a signaling system that results in light perception and vision

Retina

Eye

Description

Proteins have complex shapes that can be permanently altered. **Denaturation** of a protein is like unraveling its three-dimensional shape. Imagine that you take a necklace of beads on a string and wad it into a ball. The ball represents the normal protein shape. If you let the necklace return to its original loop, it resembles what happens during denaturing. The forces that maintain the secondary, tertiary, and quaternary structures (three-dimensional shape) of the protein are broken. Just as the beads remain together on the strand, it's important to note that the amino acids remain joined together, as denaturing has no effect on the covalent bonds (**peptide bonds**) linking amino acids. In other words, you still have a necklace after denaturing.

Form often follows function in biomolecules such as proteins. Therefore, when a protein loses its three-dimensional shape, it typically loses its function. For example, the acidic pH of your stomach's gastric juice denatures the proteins in foods that you ingest. This is a normal part of the digestive process.

Forces Holding Proteins Together

Let's examine four different forces that maintain the normal three-dimensional shapes of proteins.

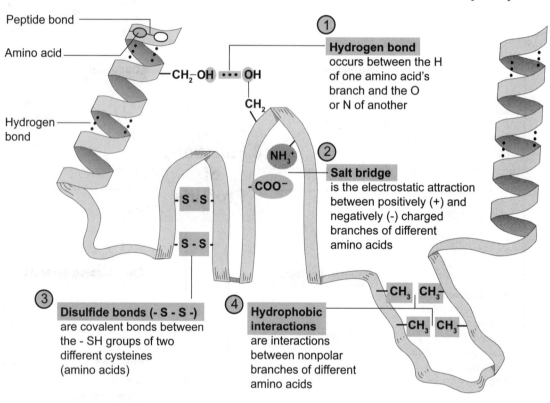

① **Hydrogen bond** occurs between the H of one amino acid's branch and the O or N of another

Peptide bond

Amino acid

Hydrogen bond

—CH₂–OH ••• OH

② **Salt bridge** is the electrostatic attraction between positively (+) and negatively (-) charged branches of different amino acids

③ **Disulfide bonds (- S - S -)** are covalent bonds between the - SH groups of two different cysteines (amino acids)

④ **Hydrophobic interactions** are interactions between nonpolar branches of different amino acids

Denatured Protein

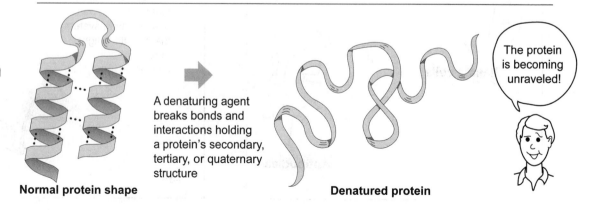

A denaturing agent breaks bonds and interactions holding a protein's secondary, tertiary, or quaternary structure

Normal protein shape

Denatured protein

The protein is becoming unraveled!

**Agents
That
Cause
Denaturation**

Denaturing agent	Interactions/bonds broken
Heat	Hydrogen bonds
Acids and bases	Salt bridges, hydrogen bonds
Alcohols	Hydrophobic interactions
Heavy metal ions (*Example:* Hg^{+2}, Pb^{+2})	Disulfide bonds

Application #1: Frying an egg

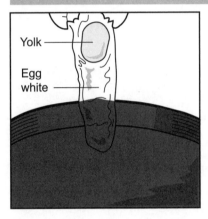

A fresh egg has a yellow yolk and a translucent egg white. The protein is found in the egg white.

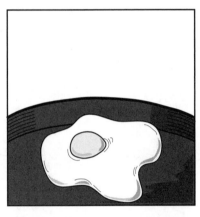

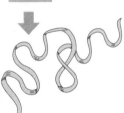

Denaturing agent:

HEAT

Denatured protein

As the egg white is heated, the protein in it is denatured. Evidence for this is that it becomes firmer and whiter in color.

Application #2: Making yogurt

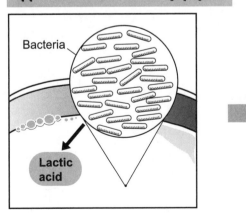

In the process of making yogurt, bacteria added to milk produce lactic acid.

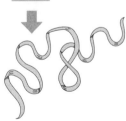

Denaturing agent:

ACID

Denatured protein

The lactic acid lowers the pH of the milk. This change in pH denatures the protein in the milk to form yogurt.

Enzymes

Description

Enzymes are proteins that act as **catalysts** to increase the rate of chemical reactions. Astonishingly, some raise the reaction rate by as much as one million times! Without enzymes, the chemical reactions needed to sustain life would not be possible. Consequently, they are found throughout your body in places, including the plasma, the digestive tract, and inside your body cells.

The three-dimensional shape of an enzyme is vital to its function. Every enzyme has a cleft or depression called the **active site**. This is where it binds the **substrate**, the substance upon which the enzyme acts. Every enzyme has a shape that specifically fits with only one kind of substrate. Digestive enzymes are common, so they are a good example to use. Every time you eat food, enzyme-producing cells either lining your digestive tract or within glands secrete many different digestive enzymes to assist in breaking down food into smaller substances that can be used by body cells.

Lactose is a disaccharide, or "double sugar," that breaks down into its two component sugars, galactose and glucose. Imagine that you just ate some yogurt for breakfast. It contains many nutrients, but let's focus only on the lactose. After swallowing the yogurt, it moves down your esophagus, into your stomach, then into your small intestine. Enzyme-producing cells lining the small intestine secrete the enzyme **lactase**. As lactase mixes with the food inside the intestine, the lactase comes in contact with the lactose, and the two bind together. This is called the **enzyme–substrate complex**. Next, a type of chemical reaction called a **hydrolysis** reaction occurs in which a water molecule helps to break the covalent bond between the two sugars. The two sugars, glucose and galactose, are then released from the enzyme, absorbed into the blood, and delivered to body cells. Once inside body cells, they can either be used as a fuel immediately or stored for later use.

Have you heard of lactose intolerance? This is the condition in which a person is unable to digest lactose. As you might suspect, the cause of this disorder is a deficiency in the production of the enzyme lactase. To compensate for this problem, lactase can be purchased from stores and added to dairy products prior to ingestion. Some dairy products such as milk claim to be "lactose-free." This indicates that milk already has been treated with lactase and that the lactose already has been digested to galactose and glucose.

Two Models for Enzyme–Substrate Interaction

Two models commonly are used to describe how the substrate binds to the enzyme: (1) **lock and key model**, and (2) **induced fit model**.

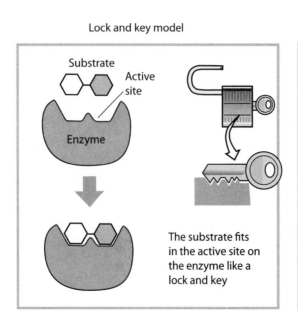

Lock and key model

The substrate fits in the active site on the enzyme like a lock and key

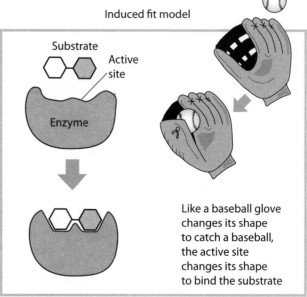

Induced fit model

Like a baseball glove changes its shape to catch a baseball, the active site changes its shape to bind the substrate

Glucose

Galactose

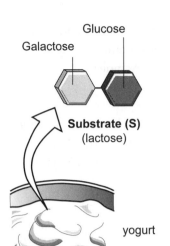

Substrate (S)
(lactose)

yogurt

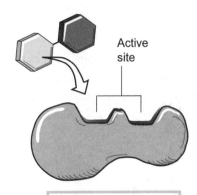

Active
site

1 **Lactose** is the sugar
found in dairy products
such as milk and yogurt.
We break it down in our
small intestine into its
two component sugars—
galactose and glucose.

2 **Enzyme (E)**
(lactase)

The enzyme lactase
is made by cells in
your small intestine
and digests *only* the
substrate lactose.

H_2O

Covalent
bond

**ENZYME
FUNCTION**
(*Example:* digesting
the sugar lactose
with the enzyme
lactase)

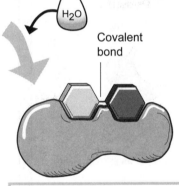

5 **Free enzyme**

The lactase enzyme
is now free to bind
another lactose molecule
and repeat the cycle.

3 **Enzyme-substrate (E•S)
complex**

Inside the small intestine,
lactose binds to the active
site on lactase. Then, with
the help of a water molecule,
the covalent bond shown is
broken, forming the products
glucose and galactose.

Galactose

Glucose

4 **Products formed**

The products galactose and
glucose are released from
the enzyme. These sugars
are small enough to be
absorbed into the blood
and delivered to body cells
where they can be used as
fuel or stored.

Description

The three key factors affecting enzyme activity are pH, temperature, and substrate concentration. For each of these variables there is an optimal value where maximum enzyme activity is reached. Digestive enzymes are common, so they will often be used here as examples.

pH

Different enzymes have different optimal pH values at which maximum enzyme activity is reached. Changing the pH away from any given optimal level can destroy the enzyme's function by denaturing it. Let's consider the optimal pH for three different digestive enzymes: pepsin, sucrase, and trypsin.

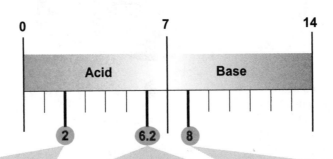

The pH scale ranges from 0 - 14, with 7 being the neutral point. Anything less than 7 is an acid; anything greater than 7 is a base.

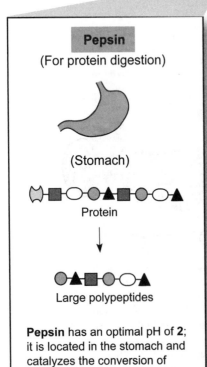

Pepsin

(For protein digestion)

(Stomach)

Protein

Large polypeptides

Pepsin has an optimal pH of **2**; it is located in the stomach and catalyzes the conversion of proteins into large polypeptides.

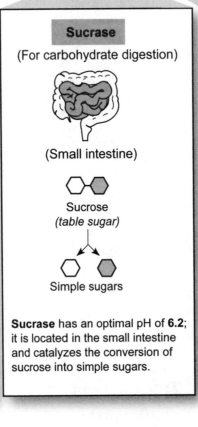

Sucrase

(For carbohydrate digestion)

(Small intestine)

Sucrose
(table sugar)

Simple sugars

Sucrase has an optimal pH of **6.2**; it is located in the small intestine and catalyzes the conversion of sucrose into simple sugars.

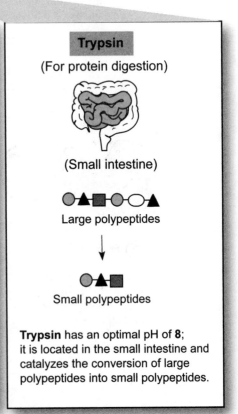

Trypsin

(For protein digestion)

(Small intestine)

Large polypeptides

Small polypeptides

Trypsin has an optimal pH of **8**; it is located in the small intestine and catalyzes the conversion of large polypeptides into small polypeptides.

Temperature

All enzymes are affected by changes in temperature. Here are some general guidelines regarding how they are affected.

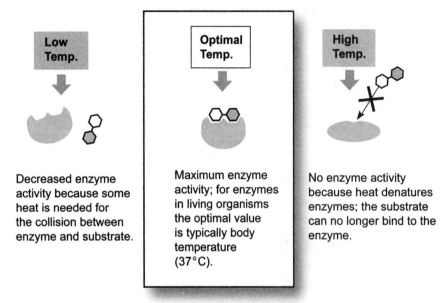

Low Temp.

Decreased enzyme activity because some heat is needed for the collision between enzyme and substrate.

Optimal Temp.

Maximum enzyme activity; for enzymes in living organisms the optimal value is typically body temperature (37°C).

High Temp.

No enzyme activity because heat denatures enzymes; the substrate can no longer bind to the enzyme.

Substrate Concentration

Consider the enzyme sucrase working on the substrate sucrose (table sugar) in the small intestine. The concentration of this substrate fluctuates depending on how much sugar you ingest. The graph below assumes that the enzyme concentration remains constant and shows that as substrate concentration increases, enzyme activity also increases, until it plateaus.

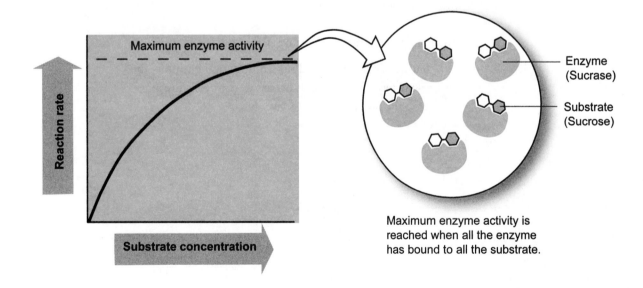

Maximum enzyme activity

Reaction rate

Substrate concentration

Enzyme (Sucrase)

Substrate (Sucrose)

Maximum enzyme activity is reached when all the enzyme has bound to all the substrate.

Description

Inhibitors are molecules that cause an enzyme to lose its normal function. We will discuss two types of inhibitors: (1) **competitive inhibitors**, and (2) **noncompetitive inhibitors**. Both types usually inhibit the enzyme temporarily—not permanently. Their effects on the enzyme last only as long as they are bound to it.

Competitive Inhibitors

A competitive inhibitor gets its name from the fact that it competes with the substrate for the same binding site on the enzyme. This means that the substrate and the inhibitor share a similar structure and have a shape specific to the active site. By preventing the substrate from binding, the enzyme cannot catalyze the reaction. This means that if the concentration of inhibitor is greater than the concentration of substrate, the inhibitor will bind more often and win the competition. Alternatively, if the concentration of substrate increases substantially, it can outcompete the inhibitor and the enzyme function is restored.

> An example of a competitive inhibitor is:
>
> - **Penicillin:** This antibiotic inhibits an enzyme needed for formation of the bacterial cell wall. Without this vital structure, the bacteria die.

Noncompetitive Inhibitors

A noncompetitive inhibitor gets its name from the fact that it does not compete with the substrate for the same binding site on the enzyme. Instead, it has a different binding site at another location on the enzyme. This means that it's structurally different from the substrate. The enzyme can be bound to the substrate, the inhibitor, or both. As long as the inhibitor is bound, however, the reaction cannot be catalyzed. Increasing the concentration of substrate will not matter in this case because the inhibitor has a unique binding site.

> An example of a noncompetitive inhibitor is:
>
> - **Alanine:** This amino acid noncompetitively inhibits the enzyme pyruvate kinase; it occurs in your cells during metabolic processes such as glycolysis.

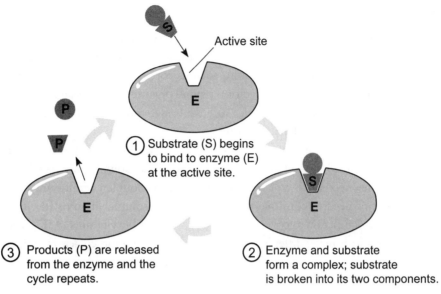

① Substrate (S) begins to bind to enzyme (E) at the active site.

② Enzyme and substrate form a complex; substrate is broken into its two components.

③ Products (P) are released from the enzyme and the cycle repeats.

Competitive inhibitors

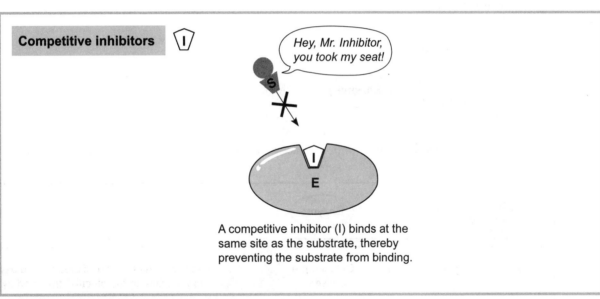

A competitive inhibitor (I) binds at the same site as the substrate, thereby preventing the substrate from binding.

Noncompetitive inhibitors

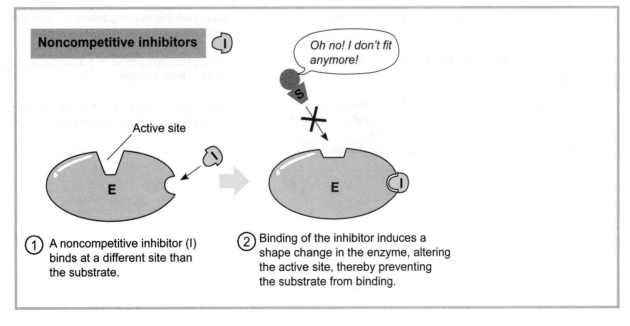

① A noncompetitive inhibitor (I) binds at a different site than the substrate.

② Binding of the inhibitor induces a shape change in the enzyme, altering the active site, thereby preventing the substrate from binding.

Description

Some enzymes catalyze reactions by themselves, and others require the help of a specific **cofactor**. If the cofactor is not present, the enzyme will not be able to function. Cofactors take different chemical forms. Examples are metal ions such as copper (Cu^{+2}), iron (Fe^{+2} or Fe^{+3}), magnesium (Mg^{+2}), and zinc (Zn^{+2}). When the cofactor is a small organic molecule such as vitamin C, it is given a more specific name—a coenzyme. To make sure we understand the difference in terminology, a coenzyme is one specific type of cofactor. Got it?

Some vitamins function as coenzymes, so we need to explore them more. Vitamins are divided into two broad groups: (1) water-soluble, and (2) fat-soluble. Water-soluble vitamins are lost in the urine, so they need to be replaced daily, whereas fat-souble vitamins are stored longer in body tissues, so we do not need to consume them daily. Water-soluble vitamins (B vitamins, vitamin C, folic acid, and biotin), function as coenzymes, whereas fat-soluble vitamins (A, D, E, and K) do not. Coenzymes often work with many different enzymes, whereas metal ion cofactors typically work with a specific enzyme. Let's examine metal ions as cofactors and coenzymes as cofactors.

Metal Ions as Cofactors

This represents the concept of metal ions as cofactors:

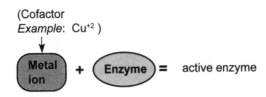

(Cofactor
Example: Cu^{+2})

Metal ion + Enzyme = active enzyme

The table below lists four common cofactors, their associated enzymes, and their respective functions:

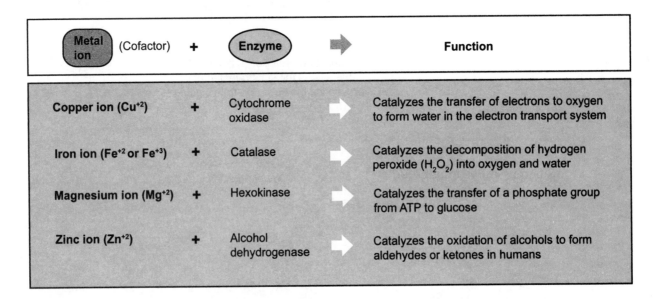

Metal ion (Cofactor)	+	Enzyme		Function
Copper ion (Cu^{+2})	+	Cytochrome oxidase	➡	Catalyzes the transfer of electrons to oxygen to form water in the electron transport system
Iron ion (Fe^{+2} or Fe^{+3})	+	Catalase	➡	Catalyzes the decomposition of hydrogen peroxide (H_2O_2) into oxygen and water
Magnesium ion (Mg^{+2})	+	Hexokinase	➡	Catalyzes the transfer of a phosphate group from ATP to glucose
Zinc ion (Zn^{+2})	+	Alcohol dehydrogenase	➡	Catalyzes the oxidation of alcohols to form aldehydes or ketones in humans

Coenzymes as Cofactors

This represents the coenzyme concept:

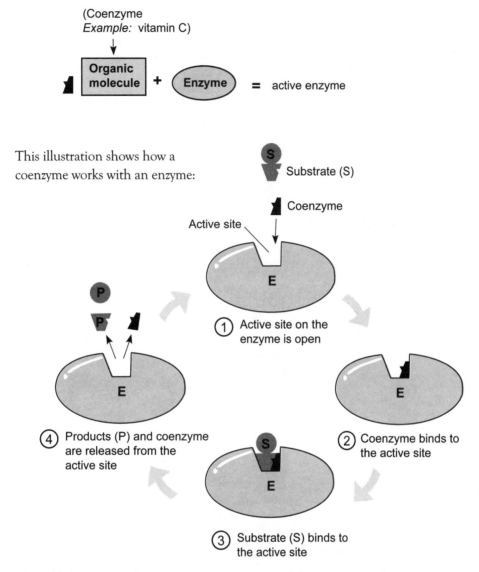

This illustration shows how a coenzyme works with an enzyme:

① Active site on the enzyme is open

② Coenzyme binds to the active site

③ Substrate (S) binds to the active site

④ Products (P) and coenzyme are released from the active site

The table below lists four common **coenzymes** and their respective functions:

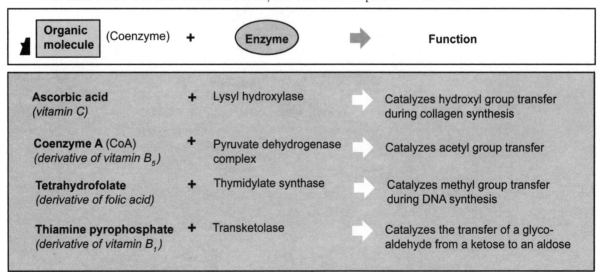

Organic molecule (Coenzyme) +	Enzyme	Function
Ascorbic acid (*vitamin C*) +	Lysyl hydroxylase	Catalyzes hydroxyl group transfer during collagen synthesis
Coenzyme A (CoA) (*derivative of vitamin B_5*) +	Pyruvate dehydrogenase complex	Catalyzes acetyl group transfer
Tetrahydrofolate (*derivative of folic acid*) +	Thymidylate synthase	Catalyzes methyl group transfer during DNA synthesis
Thiamine pyrophosphate (*derivative of vitamin B_1*) +	Transketolase	Catalyzes the transfer of a glyco-aldehyde from a ketose to an aldose

Carbohydrates

CARBOHYDRATES

Description

When runners talk about "carb loading" or advertisers talk about a "low-carb diet," they are referring to carbohydrates. In simplest terms, carbohydrates are "sugars," or polymers made from sugars that often have names ending in the suffix –*ose*, such as glucose. They are composed of the elements carbon, hydrogen, and oxygen. From simplest to most complex, carbohydrates are divided into the following three major types:

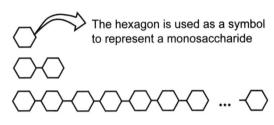

The hexagon is used as a symbol to represent a monosaccharide

- **Monosaccharides** (single sugars)
  Example: glucose (blood sugar)
- **Disaccharides** (double sugars)
  Example: maltose (grain sugar)
- **Polysaccharides** (many sugars)
  Example: amylose (plant starch)

Carbohydrates have several functions vital to all living organisms. As an overview, let's put them into three functional categories: energy, stored energy, and structural components.

Major Functional Types

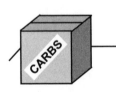

These three functional categories are presented visually on the facing page.

① **Energy:** Like gasoline for your car, monosaccharides are used as energy for body cells.

Example:

- **Glucose** ("blood sugar"): The oxidation of the monosaccharide glucose within body cells produces carbon dioxide and water. During this process, free energy is released that can be used to power cellular work such as muscle contraction.

② **Structural components:** Some serve as markers on cell surfaces; others are key parts of molecular structures.

Examples:

- **Cellulose:** This polysaccharide is found in plants and is composed of long, straight chains of glucose molecules. It makes the cell walls around plant cells rigid.
- **DNA:** The backbone of the DNA double helix is made of deoxyribose sugars and phosphate groups.
- **Antigens for ABO blood groups:** Your blood type (A, B, AB, or O) is determined by proteins on the surface of red blood cells. Attached to these proteins are short-chain carbohydrates.

③ **Stored energy:** Carbohydrates are held in reserve like items in a storage box.

Examples:

- **Glycogen:** This polysaccharide is found in animal tissues (such as liver and muscle cells) and is composed of long chains of glucose molecules. When we ingest excess glucose that is not immediately used for energy, glycogen is produced.
- **Amylose** (type of starch): This polysaccharide is found in plants and is composed of a long chain of glucose molecules that functions as an important energy storage molecule.

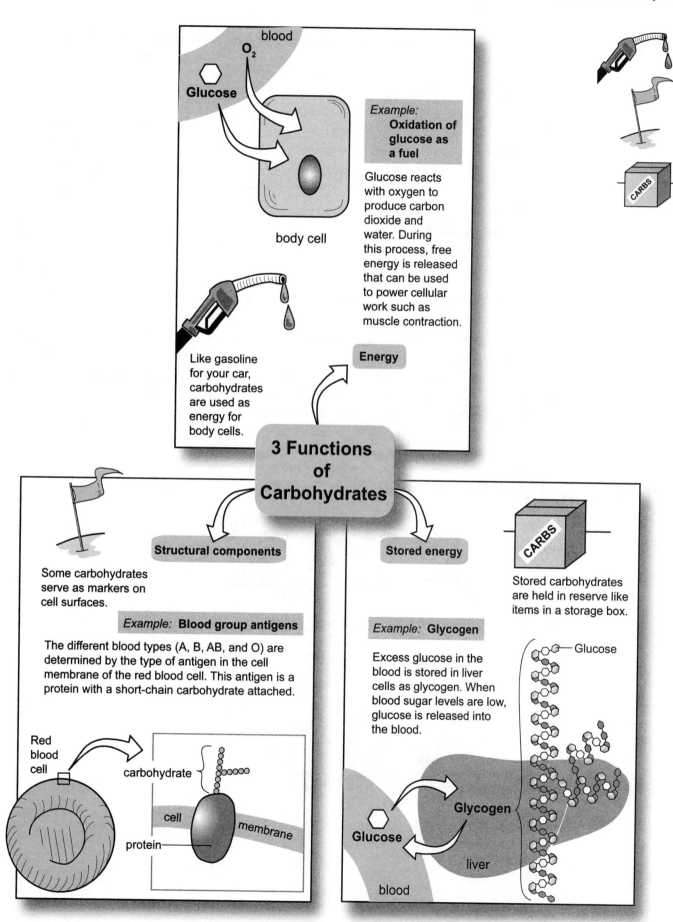

blood

O₂

Glucose

body cell

Example:
Oxidation of glucose as a fuel

Glucose reacts with oxygen to produce carbon dioxide and water. During this process, free energy is released that can be used to power cellular work such as muscle contraction.

Energy

Like gasoline for your car, carbohydrates are used as energy for body cells.

CARBS

3 Functions of Carbohydrates

Structural components

Some carbohydrates serve as markers on cell surfaces.

Example: **Blood group antigens**

The different blood types (A, B, AB, and O) are determined by the type of antigen in the cell membrane of the red blood cell. This antigen is a protein with a short-chain carbohydrate attached.

Red blood cell

carbohydrate

cell

membrane

protein

Stored energy

CARBS

Stored carbohydrates are held in reserve like items in a storage box.

Example: **Glycogen**

Excess glucose in the blood is stored in liver cells as glycogen. When blood sugar levels are low, glucose is released into the blood.

Glucose

Glycogen

Glucose

liver

blood

Description

A monosaccharide is the simplest form of carbohydrate. Monosaccharides can be either aldehydes or ketones.

- Aldehyde = *ald*ose sugar
- Ketone = *ket*ose sugar

The most common monosaccharides contain three to eight carbons.

- 3 carbons = **triose** sugar (smallest monosaccharide)
- 4 carbons = **tetrose** sugar
- 5 carbons = **pentose** sugar
- 6 carbons = **hexose** sugar
- 7 carbons = **septose** sugar
- 8 carbons = **octose** sugar (largest monosaccharide)

Remember that the prefix "mono" means "one."

Monosaccharide structures often are drawn as a straight chain of carbon atoms; however, the ring structure more accurately represents monosaccharides in aqueous solutions. A ring is formed when the carbonyl group reacts with a hydroxyl group to form a new group called the hemiacetal. Two possible mirror images, D or L, of the cyclic structure can be formed (e.g., D-glucose or L-glucose). D-glucose is the most common form found in nature.

Examples

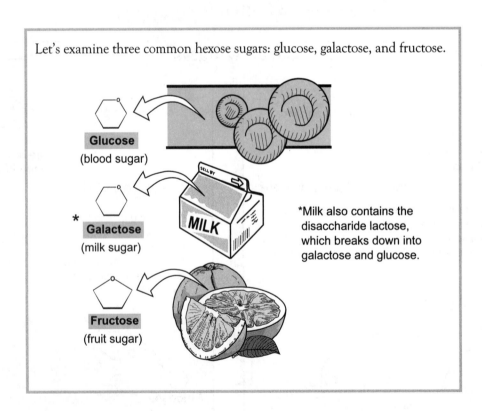

Let's examine three common hexose sugars: glucose, galactose, and fructose.

Glucose
(blood sugar)

* **Galactose**
(milk sugar)

Fructose
(fruit sugar)

*Milk also contains the disaccharide lactose, which breaks down into galactose and glucose.

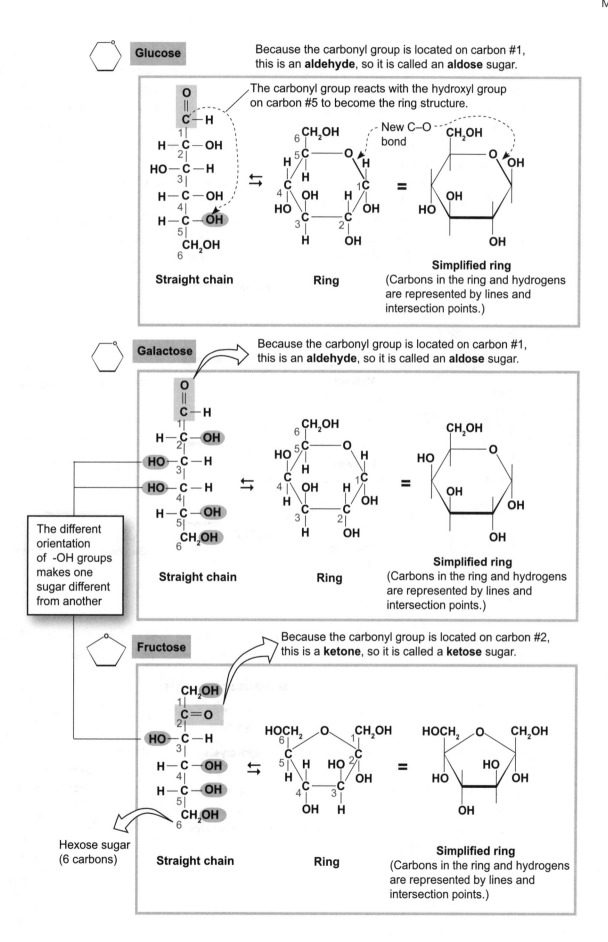

Glucose

Because the carbonyl group is located on carbon #1, this is an **aldehyde**, so it is called an **aldose** sugar.

The carbonyl group reacts with the hydroxyl group on carbon #5 to become the ring structure.

New C–O bond

Straight chain

Ring

Simplified ring
(Carbons in the ring and hydrogens are represented by lines and intersection points.)

Galactose

Because the carbonyl group is located on carbon #1, this is an **aldehyde**, so it is called an **aldose** sugar.

The different orientation of -OH groups makes one sugar different from another

Straight chain

Ring

Simplified ring
(Carbons in the ring and hydrogens are represented by lines and intersection points.)

Fructose

Because the carbonyl group is located on carbon #2, this is a **ketone**, so it is called a **ketose** sugar.

Hexose sugar
(6 carbons)

Straight chain

Ring

Simplified ring
(Carbons in the ring and hydrogens are represented by lines and intersection points.)

CARBOHYDRATES

Description

Disaccharides are "double sugars," or two monsaccharides covalently bonded together. When consumed, they have to be digested into their component monosaccharides in a process called **hydrolysis**. This occurs with the help of enzymes that catalyze reactions between the disaccharides and water molecules to yield their component monosaccharides. As an example, suppose you just ate a bowl of cereal, some yogurt, and a cup of coffee with sugar for breakfast. Without knowing it, you ingested three common disaccharides: maltose, lactose, and sucrose, respectively. The cereal was made from grain, which is a common source of maltose ("grain sugar"). The yogurt, like other dairy products such as milk and cheese, contained lactose ("milk sugar"). Last, the sugar in your coffee was sucrose ("table sugar"). Inside your small intestine, the maltose is digested to form two glucose molecules, the lactose is digested to form a glucose and a galactose, and the sucrose is digested to form glucose and fructose. These monosaccharides are small enough to travel through the blood and be transported through cell membranes and into your body's cells.

HYDROLYSIS

grains

Maltose
(grain sugar)

+ H_2O → Glucose + Glucose

yogurt

Lactose
(milk sugar)

+ H_2O → Glucose + Galactose

table sugar

Sucrose
(table sugar)

+ H_2O → Glucose + Fructose

Formation of a Disaccharide

The opposite of hydrolysis is **dehydration synthesis**, the process used to form a disaccharide. The term tells us that a new substance is made (synthesis) and it involves the loss of water (dehydration).

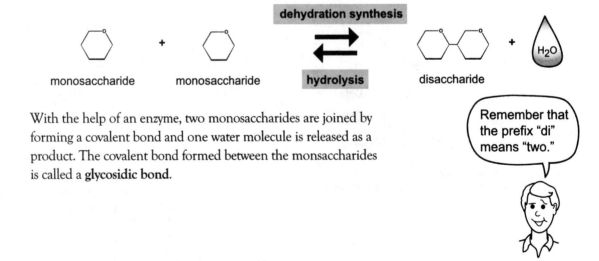

monosaccharide + monosaccharide

dehydration synthesis

⇌

hydrolysis

disaccharide + H_2O

With the help of an enzyme, two monosaccharides are joined by forming a covalent bond and one water molecule is released as a product. The covalent bond formed between the monsaccharides is called a **glycosidic bond**.

Remember that the prefix "di" means "two."

184

FORMATION of MALTOSE

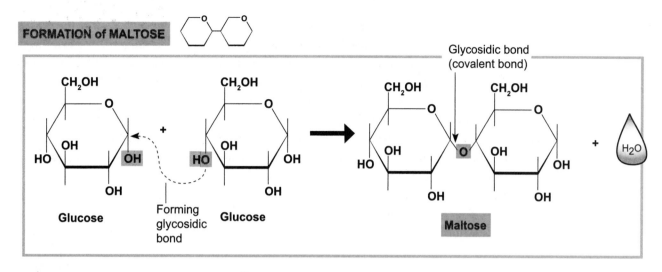

Glucose + Glucose → Maltose + H_2O

Glycosidic bond (covalent bond)

Forming glycosidic bond

FORMATION of LACTOSE

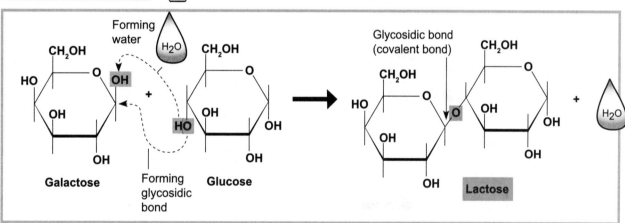

Forming water

Galactose + Glucose → Lactose + H_2O

Glycosidic bond (covalent bond)

Forming glycosidic bond

FORMATION of SUCROSE

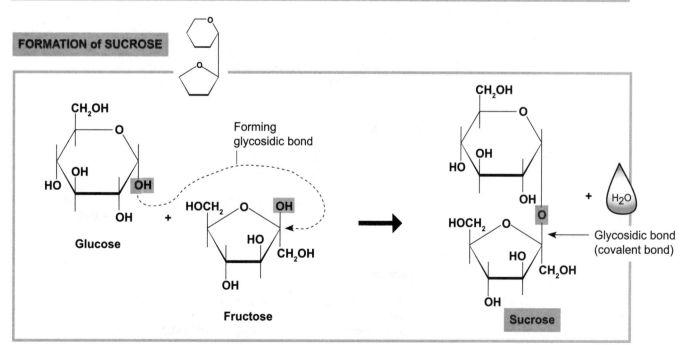

Glucose + Fructose → Sucrose + H_2O

Forming glycosidic bond

Glycosidic bond (covalent bond)

Description

Polysaccharides are long chains of monosaccharides such as glucose. Every day our bodies are making polysaccharides through a process called **dehydration synthesis** and breaking them down through an opposite process called **hydrolysis**. As an example, suppose you ate a very large lunch that included a baked potato. Part of the starch in the baked potato is a polysaccharide called **amylose**, which is a long chain of glucose molecules. Inside your digestive tract, enzymes are used along with water to break down the amylose into **oligosaccharides** (short–chain sugars) and the **disaccharide** (double sugars) maltose. Eventually, these sugars are broken down into the monosaccharide glucose and absorbed into the bloodstream. If you take a nap after lunch because you ate too much, your body will store the excess blood glucose in liver and muscle cells by converting it into another polysaccharide called **glycogen**. This process, dehydration synthesis, produces water as a product. Later, if you exercise vigorously, your blood glucose levels will begin to fall. In response, your body will use hydrolysis again to break down glycogen into glucose, thereby raising your blood glucose levels.

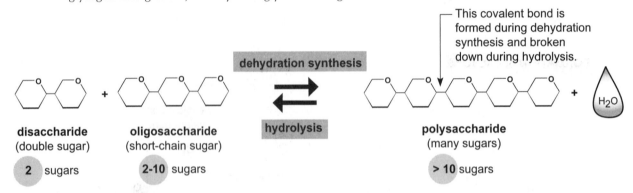

We will explore three common examples of polysaccharides based on glucose: *cellulose*, *amylose*, and *glycogen*. Illustrations of these polysaccharides are shown on the facing page.

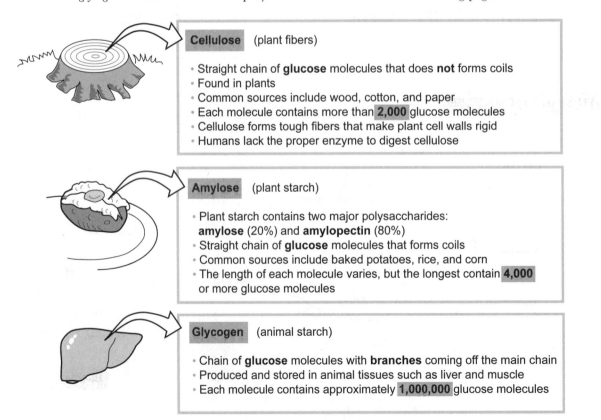

Cellulose (plant fibers)

- Straight chain of **glucose** molecules that does **not** forms coils
- Found in plants
- Common sources include wood, cotton, and paper
- Each molecule contains more than 2,000 glucose molecules
- Cellulose forms tough fibers that make plant cell walls rigid
- Humans lack the proper enzyme to digest cellulose

Amylose (plant starch)

- Plant starch contains two major polysaccharides:
  amylose (20%) and **amylopectin** (80%)
- Straight chain of **glucose** molecules that forms coils
- Common sources include baked potatoes, rice, and corn
- The length of each molecule varies, but the longest contain 4,000 or more glucose molecules

Glycogen (animal starch)

- Chain of **glucose** molecules with **branches** coming off the main chain
- Produced and stored in animal tissues such as liver and muscle
- Each molecule contains approximately 1,000,000 glucose molecules

Remember that the prefix "poly" means "many." Makes sense, right?

COMMON POLYSACCHARIDES BASED on GLUCOSE

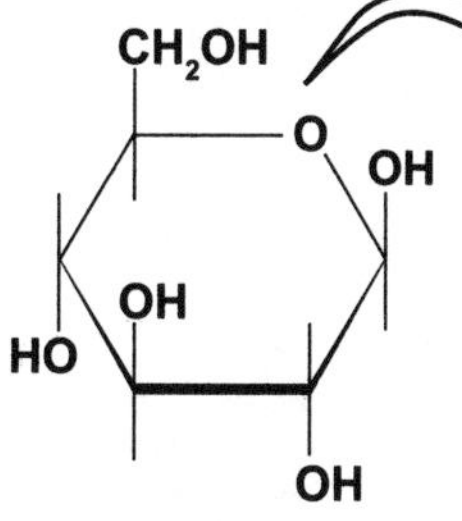

| POLYSACCHARIDE | COMMON SOURCES |

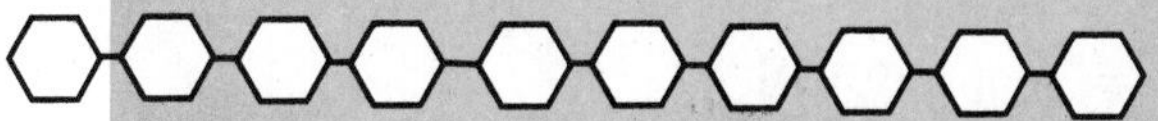

CH$_2$OH

O

OH

HO

OH

OH

Glucose

Cellulose
- Straight chain that does **not** form coils
- Found in plants

- Wood
- Cotton
- Paper

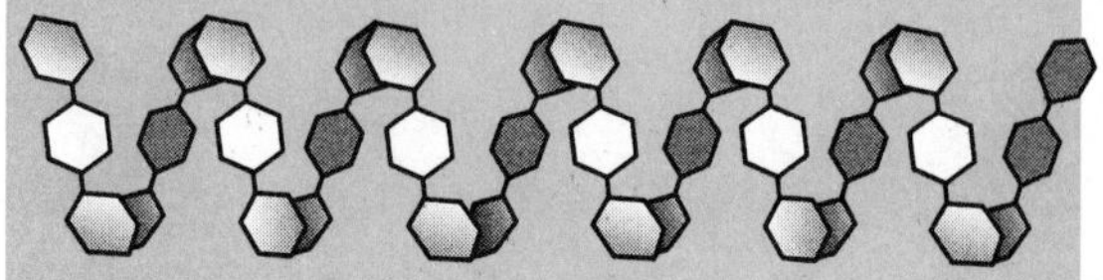

Amylose
- Straight chain that forms coils
- Produced and stored in plants

- Baked potato
- Rice
- Corn

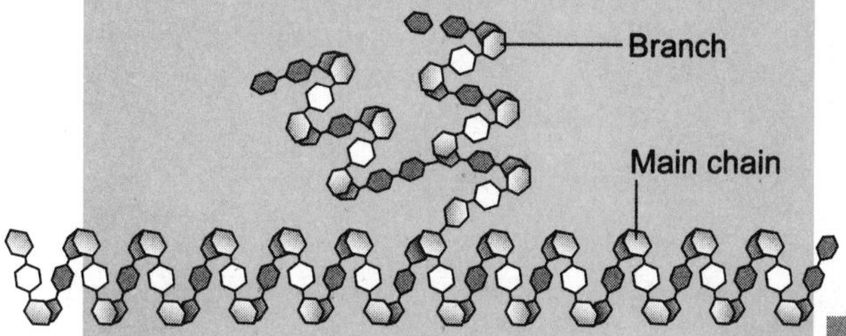

Branch

Main chain

Glycogen
- Chain with **branches** coming off main chain
- Produced and stored in animal tissues

Stored in:
- Liver
- Muscle

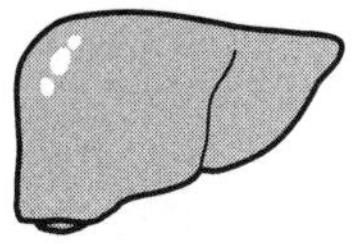

Nucleotides, Nucleic Acids, and Protein Synthesis

Description

Nucleotides are the monomers, or building blocks, for making long-chained nucleic acid polymers such as **DNA** and **RNA**. If you ate a spinach salad for lunch, you ingested DNA from inside the nucleus of the plant cells. In your digestive system, DNA is broken down into nucleotides, which then are transported through the blood to all your cells. Once inside your cells, the free nucleotides float around like raw materials at a construction site. These nucleotides are used to synthesize new DNA in your cells. A DNA nucleotide from a plant or an animal is chemically the same, so the original source doesn't matter. It just reminds you that you really are what you eat! Let's examine in more detail how a DNA nucleotide is used as a building block for making DNA.

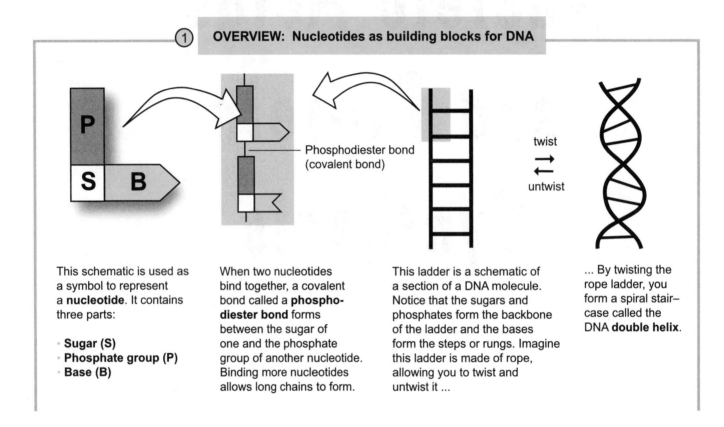

① OVERVIEW: Nucleotides as building blocks for DNA

Phosphodiester bond (covalent bond)

twist ⇄ untwist

This schematic is used as a symbol to represent a **nucleotide**. It contains three parts:

- **Sugar (S)**
- **Phosphate group (P)**
- **Base (B)**

When two nucleotides bind together, a covalent bond called a **phosphodiester bond** forms between the sugar of one and the phosphate group of another nucleotide. Binding more nucleotides allows long chains to form.

This ladder is a schematic of a section of a DNA molecule. Notice that the sugars and phosphates form the backbone of the ladder and the bases form the steps or rungs. Imagine this ladder is made of rope, allowing you to twist and untwist it ...

... By twisting the rope ladder, you form a spiral stair-case called the DNA **double helix**.

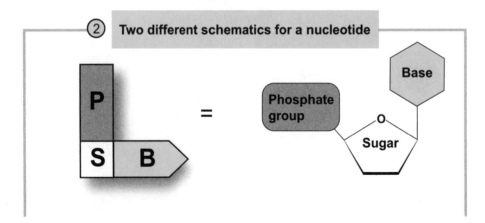

② Two different schematics for a nucleotide

P

S B =

Phosphate group

Base

Sugar

O

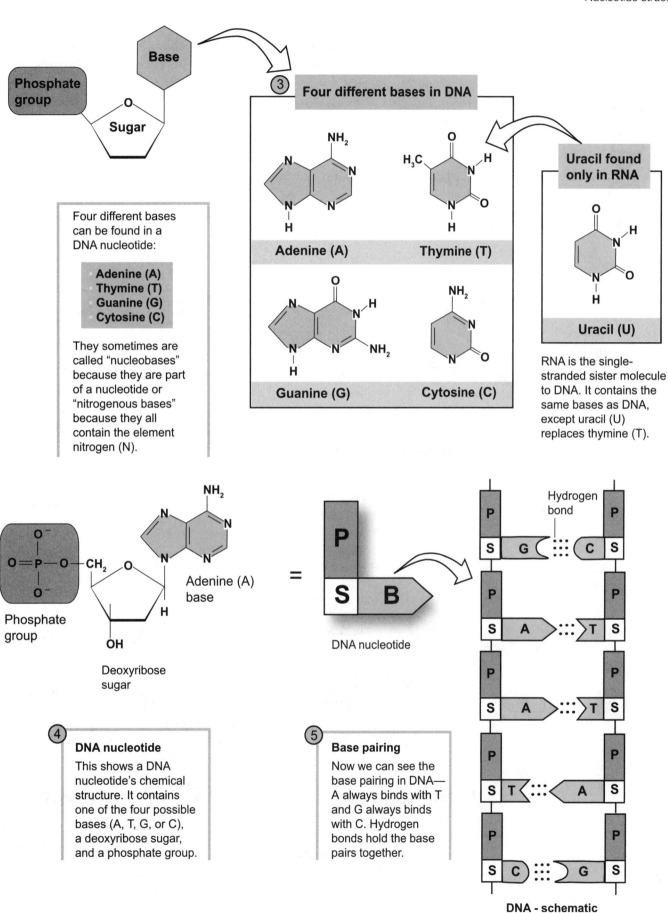

Base

Phosphate group

Sugar

③ **Four different bases in DNA**

Four different bases can be found in a DNA nucleotide:

- **Adenine (A)**
- **Thymine (T)**
- **Guanine (G)**
- **Cytosine (C)**

They sometimes are called "nucleobases" because they are part of a nucleotide or "nitrogenous bases" because they all contain the element nitrogen (N).

Adenine (A)

Thymine (T)

Guanine (G)

Cytosine (C)

Uracil found only in RNA

Uracil (U)

RNA is the single-stranded sister molecule to DNA. It contains the same bases as DNA, except uracil (U) replaces thymine (T).

Phosphate group

Adenine (A) base

Deoxyribose sugar

= **P** **S** **B**

DNA nucleotide

④ **DNA nucleotide**

This shows a DNA nucleotide's chemical structure. It contains one of the four possible bases (A, T, G, or C), a deoxyribose sugar, and a phosphate group.

⑤ **Base pairing**

Now we can see the base pairing in DNA—A always binds with T and G always binds with C. Hydrogen bonds hold the base pairs together.

Hydrogen bond

P S G ::: C S P

P S A ::: T S P

P S A ::: T S P

P S T ::: A S P

P S C ::: G S P

DNA - schematic

Description

The most common, stable form of DNA—the double helix—resembles a spiral staircase. In terms of its structure, the railing of the staircase is made of a sugar–phosphate backbone, and the steps are made of base pairs. Imagine that you cut the staircase straight down the middle through the steps. Now you have two halves of the spiral staircase. This more accurately reflects the structure of the double helix because the two halves are held together by hydrogen bonds between the base pairs.

DNA is a nucleic acid—a natural polymer made from long chains of nucleotides. It is found in the nucleus of your cells and functions as the master blueprint for making all the different proteins found throughout the body. It is tightly coiled and condensed into structures called chromosomes. Imagine that you took all the DNA in one cell and unraveled it. If you were microscopically small, guess how many steps you would have to take to climb the spiral staircase from end to end? Roughly 3 billion! I'm tired just thinking about it.

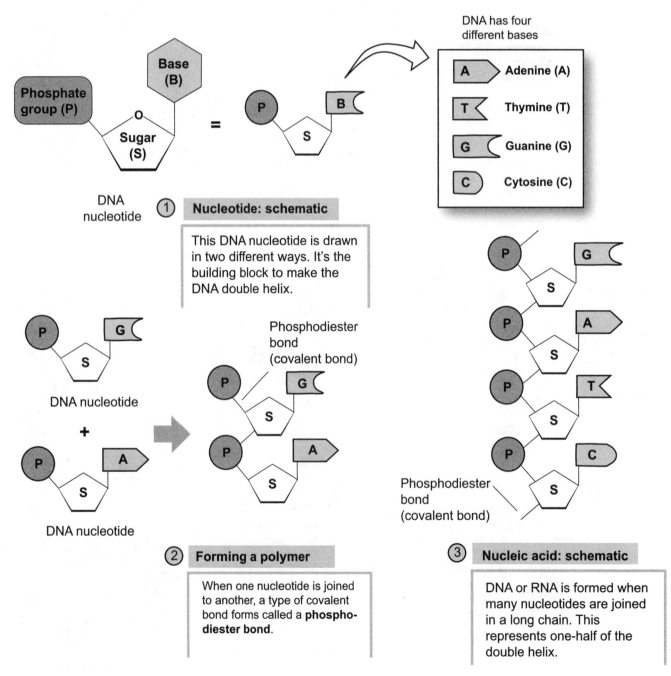

DNA has four different bases

A	Adenine (A)
T	Thymine (T)
G	Guanine (G)
C	Cytosine (C)

① Nucleotide: schematic

This DNA nucleotide is drawn in two different ways. It's the building block to make the DNA double helix.

Phosphodiester bond (covalent bond)

② Forming a polymer

When one nucleotide is joined to another, a type of covalent bond forms called a **phosphodiester bond**.

Phosphodiester bond (covalent bond)

③ Nucleic acid: schematic

DNA or RNA is formed when many nucleotides are joined in a long chain. This represents one-half of the double helix.

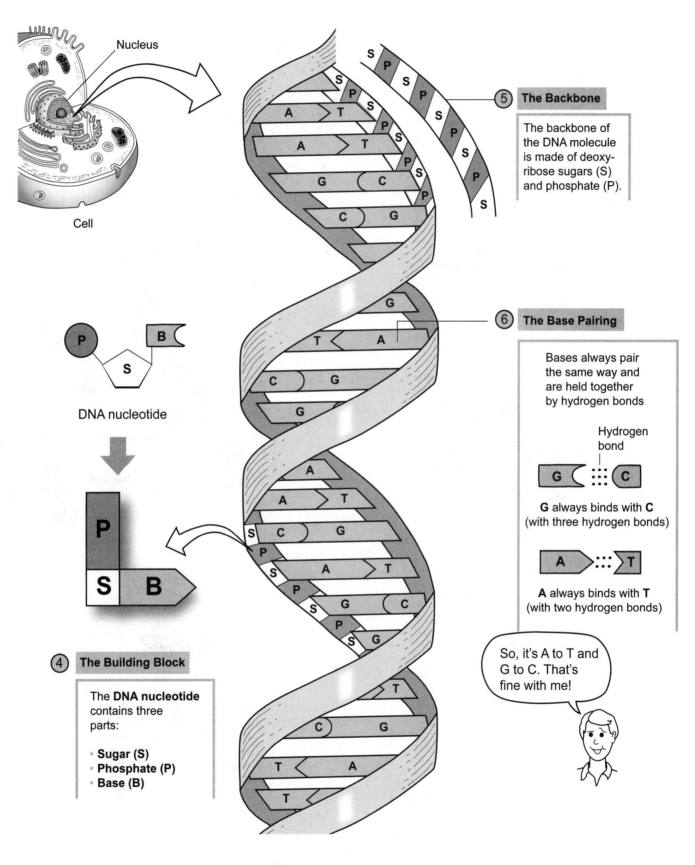

Nucleus

Cell

P · S · B

DNA nucleotide

P

S · B

⑤ **The Backbone**

The backbone of the DNA molecule is made of deoxy-ribose sugars (S) and phosphate (P).

⑥ **The Base Pairing**

Bases always pair the same way and are held together by hydrogen bonds

Hydrogen bond

G ⋯ C

G always binds with **C** (with three hydrogen bonds)

A ⋯ T

A always binds with **T** (with two hydrogen bonds)

So, it's A to T and G to C. That's fine with me!

④ **The Building Block**

The **DNA nucleotide** contains three parts:

- **Sugar (S)**
- **Phosphate (P)**
- **Base (B)**

DNA Double Helix

DNA replication

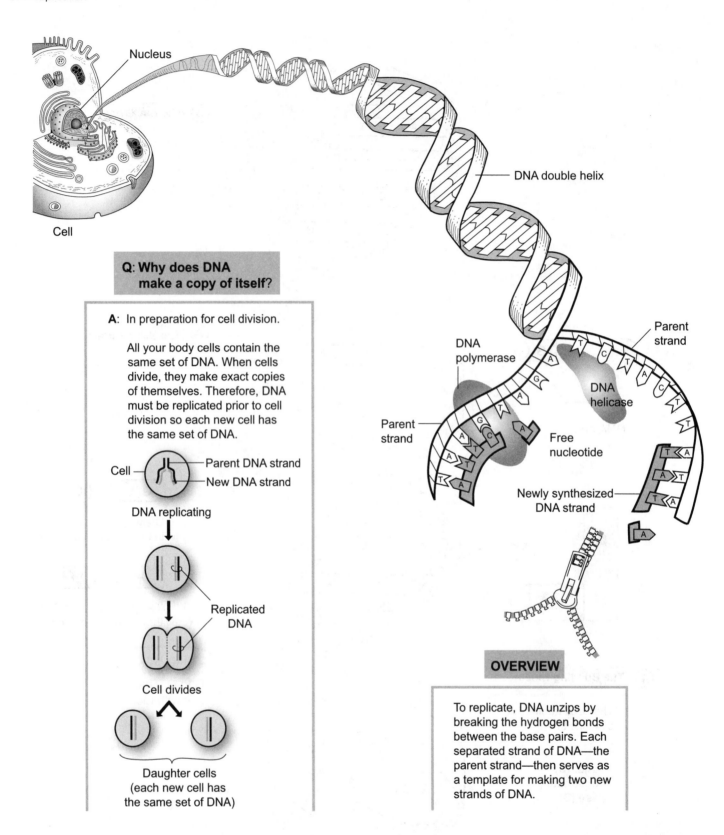

Nucleus

Cell

DNA double helix

Q: Why does DNA make a copy of itself?

A: In preparation for cell division.

All your body cells contain the same set of DNA. When cells divide, they make exact copies of themselves. Therefore, DNA must be replicated prior to cell division so each new cell has the same set of DNA.

Cell — Parent DNA strand
— New DNA strand

DNA replicating

Replicated DNA

Cell divides

Daughter cells
(each new cell has the same set of DNA)

Parent strand

DNA polymerase

DNA helicase

Parent strand

Free nucleotide

Newly synthesized DNA strand

OVERVIEW

To replicate, DNA unzips by breaking the hydrogen bonds between the base pairs. Each separated strand of DNA—the parent strand—then serves as a template for making two new strands of DNA.

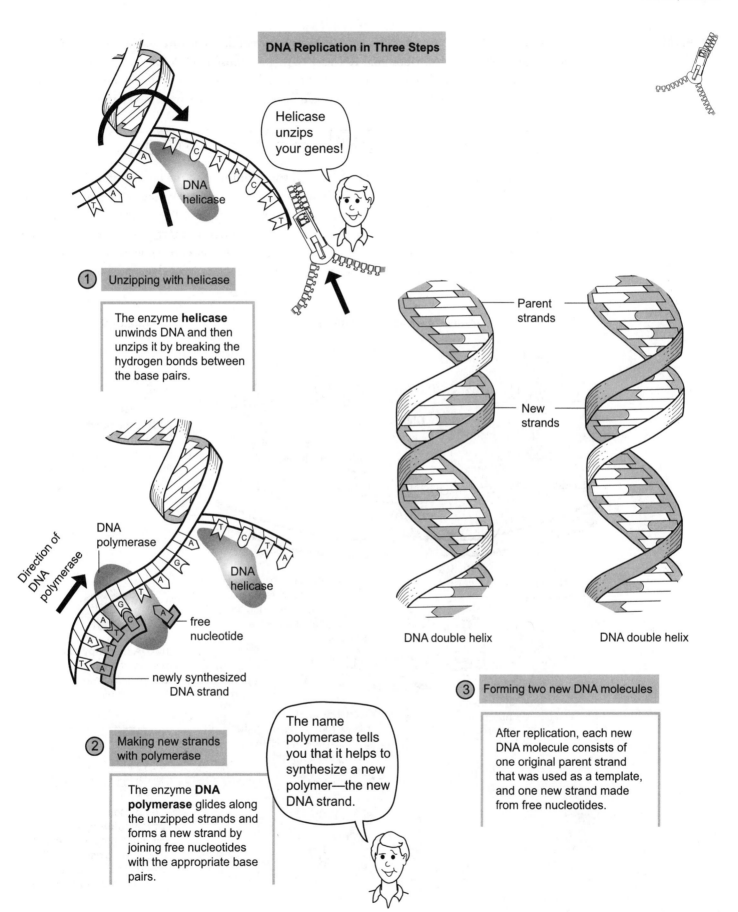

DNA Replication in Three Steps

Helicase unzips your genes!

DNA helicase

① Unzipping with helicase

The enzyme **helicase** unwinds DNA and then unzips it by breaking the hydrogen bonds between the base pairs.

Direction of DNA polymerase

DNA polymerase

DNA helicase

free nucleotide

newly synthesized DNA strand

② Making new strands with polymerase

The enzyme **DNA polymerase** glides along the unzipped strands and forms a new strand by joining free nucleotides with the appropriate base pairs.

The name polymerase tells you that it helps to synthesize a new polymer—the new DNA strand.

Parent strands

New strands

DNA double helix   DNA double helix

③ Forming two new DNA molecules

After replication, each new DNA molecule consists of one original parent strand that was used as a template, and one new strand made from free nucleotides.

Description

Like DNA, RNA is a type of nucleic acid made from nucleotides. The three major forms of RNA are: **messenger RNA** (mRNA), **ribosomal RNA** (rRNA), and **transfer RNA** (tRNA), which are best understood in the context of protein synthesis. In biological systems, DNA is used as a template to make mRNA, rRNA, and tRNA. mRNA is read as a template to determine a protein's amino acid sequence.

DNA $\xrightarrow{\text{transcription}}$ mRNA $\xrightarrow{\text{translation}}$ PROTEIN

The process of making mRNA from DNA is called **transcription** and the process of making proteins from mRNA is called **translation**. Think about the meaning of these terms. When you transcribe what someone said, you copy it in a written form. Similarly, mRNA is like a "written" form made of RNA nucleotides from copying the sequence of DNA nucleotides. In short, one nucleic acid is made from the other. In the second process, just as an interpreter translates one language into another, the nucleotide "language" of mRNA must be translated into the amino acid "language" of proteins.

Overview of Protein Synthesis

DNA

① **Transcription:** mRNA is made from the DNA template.

Nucleus

mRNA

Cytoplasm

Nuclear membrane

Nuclear pore

② **mRNA travels:** This single-stranded molecule is small enough to move through the nuclear pore and dock at the ribosome.

mRNA

Functional ribosome (rRNA)

③ **Translation:** A protein is made from interpreting the genetic code in mRNA by linking the appropriate amino acids.

Amino acid

Protein

Analogy

DNA is the **master blueprint** made of DNA nucleotides. It contains the instructions for making proteins.

mRNA is a **copy** of a portion of the master blueprint "written" using RNA nucleotides.

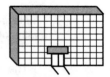

The **protein** is the **new office building** that is created from the blueprint. The steel, glass, and other raw materials used to make it are like the **amino acids**.

The RNA Players: mRNA, tRNA, and rRNA

Messenger RNA (mRNA)

The **mRNA** is a copy of a portion of the DNA master blueprint.

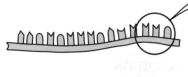

The **mRNA** is a single-stranded chain of RNA nucleotides that contain four different bases:

- **Adenine (A)**
- **Guanine (G)**
- **Cytosine (C)**
- **Uracil (U)**

Three bases form a **codon**. The three matching bases in a tRNA that bind to the codon are called the **anticodon**.

Transfer RNA (tRNA)

The **tRNA** is like a taxicab that transports its amino acid passenger to the ribosome.

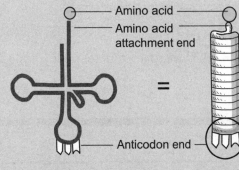

Amino acid

Amino acid attachment end

Anticodon end

The **tRNA** actually has a clover-like shape but it has been illustrated as a coil for simplicity.

The **tRNA** has two ends:
(1) Amino acid attachment end
(2) **Anticodon** end with three bases. It attaches to the mRNA codon.

Ribosomal RNA (rRNA)

The ribosome is like the **protein factory** because it is the site where proteins are made.

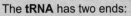

Large subunit

Small subunit

Subunits attach

Subunits detach

Functional ribosome

P site A site

The large subunit has binding sites—P and A—for tRNA. The large and small subunits attach to form a functional ribosome.

Description

The **genetic code** is a three–base sequence in which each triplet corresponds to a specific **amino acid**. The triplets—called **codons**—are located on messenger RNA (mRNA). For example, the codon UUU encodes for the amino acid phenylalanine (Phe). Each letter "U" corresponds to the base "uracil." This coding process determines protein structure in living organisms by controlling the sequence of amino acids that are bonded together. This is part of the larger process of **protein synthesis**, so take a look at that process to get a glimpse of the bigger picture (see p. 200).

Analogy: Morse Code

As an analogy, let's compare the genetic code to Morse code.

	CODE	TRANSLATION
Morse code	··─· ··─ ─··	FUN
Genetic code	UUU	The amino acid, **phenylalanine (Phe)**

Morse code uses dots and dashes to represent each letter in the alphabet. Just as Morse code translates dots and dashes into words, the genetic code translates the "language" of nucleotide bases into the "language" of amino acids.

How the Code Works

For example, the **codon UUU** translates into the amino acid **phenylalanine (Phe)**.

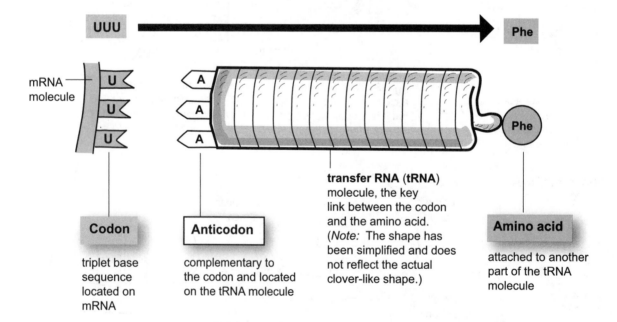

UUU ──────────────▶ Phe

mRNA molecule

Codon

triplet base sequence located on mRNA

Anticodon

complementary to the codon and located on the tRNA molecule

transfer RNA (tRNA) molecule, the key link between the codon and the amino acid. (*Note:* The shape has been simplified and does not reflect the actual clover-like shape.)

Amino acid

attached to another part of the tRNA molecule

To better understand how tRNA works in the process of protein synthesis, see p. 200.

The Genetic Code

The table below lists all of the 64 possible codons with their corresponding amino acids. The "start" codon indicates where the translation process begins, and a "stop" codon indicates where it ends.

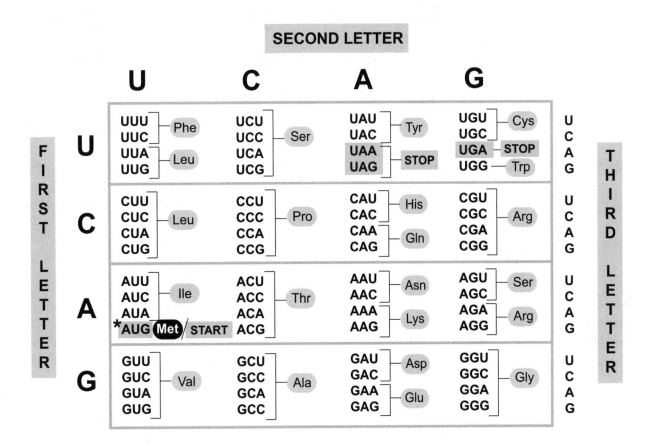

NOTE: For the full name of each amino acid, see p. 158.

✱ The "start" codon

Practice Problems

Using the table above, answer the following questions:

1 Identify the "START" codon.

2 Identify the three different "STOP" codons.

3 How many different codons does the amino acid proline (Pro) have?

4 The codon ACC translates into which amino acid?

Answers

Overview

Protein synthesis is a major function in most cells. It requires that DNA first be transcribed into a similar molecule called messenger RNA, or mRNA. Then the mRNA must be translated from the nucleotide "language" of RNA into the amino acid "language" of proteins.

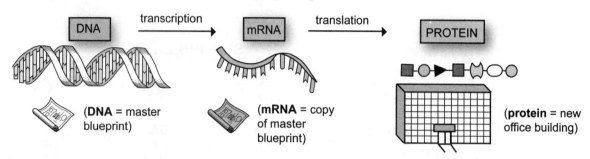

(**DNA** = master blueprint)

(**mRNA** = copy of master blueprint)

(**protein** = new office building)

Analogy

The process of protein synthesis in a cell is like constructing a new office building. The cytoplasm is like the construction site, and the DNA is like the master blueprint. The master blueprint for the office building is stored in the trailer on the construction site, just as DNA is stored in the nucleus (as it is too large to leave). Just as copies can be made of the original blueprint, a copy of DNA is made in the form of mRNA. The mRNA is small enough to leave the nucleus, enter the cytoplasm, and bind to the ribosomes, which are like the construction workers. The raw materials for the office building are steel, concrete, and glass, to name a few. To build a protein, the needed raw materials are amino acids.

Basic Steps in Protein Synthesis

① **Transcription:** The enzyme RNA Polymerase travels along an "unzipped" section of DNA for the purpose of making mRNA. The building blocks for making mRNA are called RNA nucleotides. The enzyme uses one side of the DNA molecule as a template to match the base of the RNA nucleotide to the corresponding base in DNA. As the RNA nucleotides are lined up next to each other, they are covalently bonded to eventually form a short, single-stranded mRNA molecule.

② **RNA Processing:** Like old film being spliced together during editing, the single-stranded mRNA molecule also is processed. Small segments called introns are removed from the molecule.

③ The single-stranded mRNA is small enough to leave the nucleus through the nuclear pore and move into the cytoplasm, where it binds to the small subunit of a ribosome.

④ A transfer RNA (tRNA) molecule has an amino acid binding site section at one end and a three-base section at the other end, called the anticodon. Just as a taxi transports passengers, the tRNA transports the proper amino acid to the ribosome. With the help of an enzyme and free energy from ATP hydrolysis, one specific amino acid is bonded to one specific tRNA molecule based on its anticodon sequence.

⑤ The mRNA message is read in groups of three bases. Each triplet group is called a codon. The tRNA with the amino acid called methionine always binds to the beginning of the message called the start codon. Next, the large ribosomal subunit binds to the small ribosomal subunit to activate the ribosome. The large subunit contains two binding sites for tRNAs called the P site and the A site.

⑥ The initial tRNA binds to the P site. One by one, amino acids are positioned next to each other, then bonded together. This results in a growing polypeptide chain. Here is how it works:

⑦ A tRNA with the anticodon matching the mRNA codon arrives at the A site with the next amino acid to be added to the chain. The growing polypeptide chain at the P site is transferred to the tRNA at the A site as it binds to the next amino acid to be added to the chain.

⑧ As the ribosome moves to the right, the tRNA moves from the A site over to the P site. Then the tRNA with the next amino acid to be added to the chain arrives at the A site and the process repeats itself.

⑨ The process of growing the polypeptide chain ends when a special codon called the stop codon is reached at the A site. The final polypeptide is released from the ribosome along with the last tRNA. Finally, the ribosome separates into its large and small subunits. This marks the end of protein synthesis.

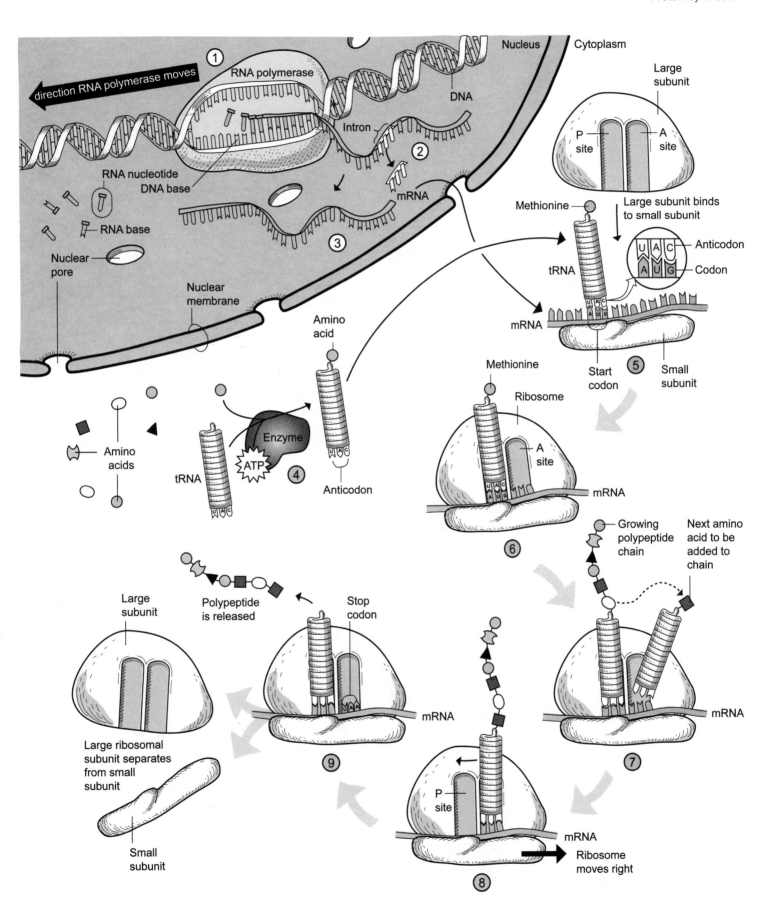

Lipids

Description

Lipids (*lipos* = fat) consist of a broad category of organic compounds that include waxes, fats and oils, glycerophospholipids, and steroids. They are made of carbon, hydrogen, and oxygen and are not soluble in water. Most lipids, except steroids, contain fatty acid chains. The fatty acid chains are composed of carbon and hydrogen ranging from 10–20 carbons in length. The total number of fatty acid chains in a lipid molecule ranges from one to three, depending on the type of lipid. For example, glycerophospholipids contain two fatty acid chains per molecule, and triacylglycerols (in fats and oils) contain three fatty acid chains per molecule.

Let's consider some practical examples. Shoe polish is a type of wax. The same is true of waxes for your car and your wood furniture. Waxes also are found in candles, soap, and some cosmetics. Fats and oils contain lipids called triacylglycerols (triglycerides). Most fats that we consume are derived from animals, and oils are derived from plants. For example, when cooking, you could choose either lard (fat) or corn oil (oil) to "grease" the frying pan. The healthier choice would be corn oil because it contains unsaturated fatty acids, which are less likely to lead to coronary artery disease. Consuming large amounts of cholesterol is often reported to be bad for your health. But what isn't mentioned is that this steroid is made in small amounts by liver cells and is a vital component in your cell membranes. To promote good health, your multivitamin contains other lipids—the fat-soluble vitamins (A, D, E, and K). Lipids are common in everyday life.

As an overview, let's compare the major classes of lipids with and without fatty acid chains.

LIPIDS with FATTY ACID CHAINS

Waxes	Found in commerical products such as shoe polish, furniture polish, car wax, candles, soaps, and cosmetics
Fats and oils (contain triacylglycerols)	Found in commerical products such as lard, corn oil, olive oil, and canola oil
Glycerophospholipids	Found in cell membranes

LIPIDS without FATTY ACID CHAINS

Steroids	Examples include sex hormones such as progesterone, estrogen, and testosterone

As an introduction to lipids, the facing page illustrates and compares the structure and function of three classes of lipids: triacylglycerols (triglycerides), glycerophospholipids, and steroids.

Comparison of three different classes of lipids: triacylglycerols, glycerophospholipids, and steroids.

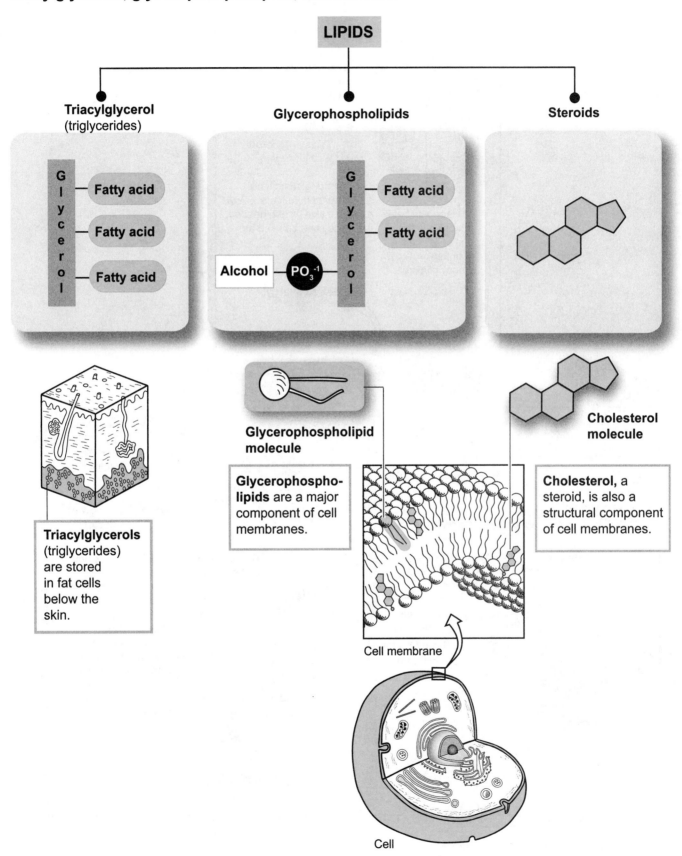

LIPIDS

Triacylglycerol (triglycerides)

Glycerol — Fatty acid
Glycerol — Fatty acid
Glycerol — Fatty acid

Glycerophospholipids

Alcohol — PO$_3^{-1}$ — Glycerol — Fatty acid
Glycerol — Fatty acid

Steroids

Triacylglycerols (triglycerides) are stored in fat cells below the skin.

Glycerophospholipid molecule

Glycerophospho-lipids are a major component of cell membranes.

Cholesterol molecule

Cholesterol, a steroid, is also a structural component of cell membranes.

Cell membrane

Cell

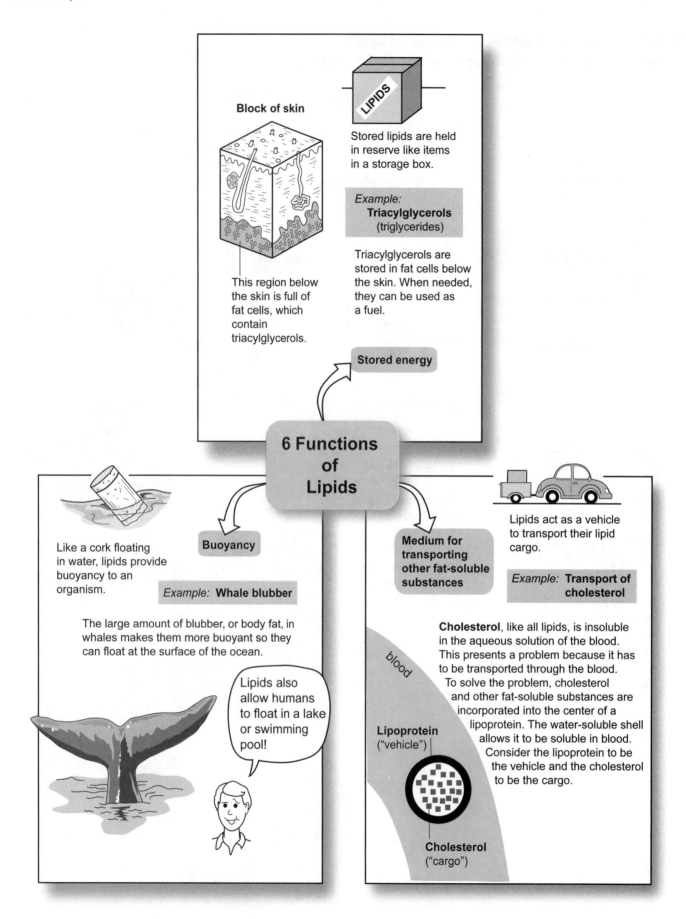

Block of skin

LIPIDS

Stored lipids are held in reserve like items in a storage box.

Example:
Triacylglycerols
(triglycerides)

Triacylglycerols are stored in fat cells below the skin. When needed, they can be used as a fuel.

This region below the skin is full of fat cells, which contain triacylglycerols.

Stored energy

6 Functions of Lipids

Like a cork floating in water, lipids provide buoyancy to an organism.

Buoyancy

Example: **Whale blubber**

The large amount of blubber, or body fat, in whales makes them more buoyant so they can float at the surface of the ocean.

Lipids also allow humans to float in a lake or swimming pool!

Medium for transporting other fat-soluble substances

Lipids act as a vehicle to transport their lipid cargo.

Example: **Transport of cholesterol**

Cholesterol, like all lipids, is insoluble in the aqueous solution of the blood. This presents a problem because it has to be transported through the blood. To solve the problem, cholesterol and other fat-soluble substances are incorporated into the center of a lipoprotein. The water-soluble shell allows it to be soluble in blood. Consider the lipoprotein to be the vehicle and the cholesterol to be the cargo.

blood

Lipoprotein
("vehicle")

Cholesterol
("cargo")

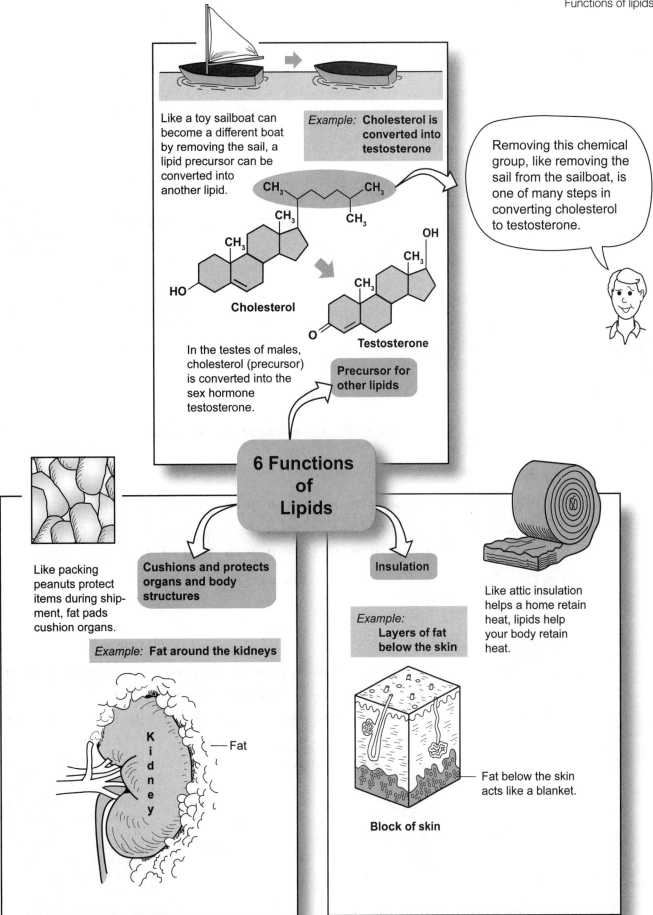

Like a toy sailboat can become a different boat by removing the sail, a lipid precursor can be converted into another lipid.

Example: **Cholesterol is converted into testosterone**

Removing this chemical group, like removing the sail from the sailboat, is one of many steps in converting cholesterol to testosterone.

Cholesterol

Testosterone

In the testes of males, cholesterol (precursor) is converted into the sex hormone testosterone.

Precursor for other lipids

6 Functions of Lipids

Like packing peanuts protect items during shipment, fat pads cushion organs.

Cushions and protects organs and body structures

Example: **Fat around the kidneys**

Kidney — Fat

Insulation

Like attic insulation helps a home retain heat, lipids help your body retain heat.

Example: **Layers of fat below the skin**

Block of skin

— Fat below the skin acts like a blanket.

Description

Triacylglycerols, commonly called **triglycerides**, are found in fats and oils. In the body, they serve as a form of stored energy in fat cells below the skin. Their name reveals a lot about their chemical structure. The prefix, "tri-" tells us that each molecule contains three fatty acid chains. The suffix, "-glycerol" identifies the name of the "backbone" of the molecule, from which the fatty acids branch.

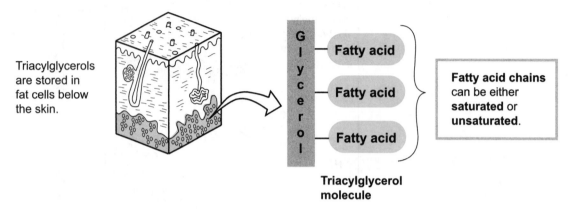

Triacylglycerols are stored in fat cells below the skin.

Fatty acid chains can be either **saturated** or **unsaturated**.

Triacylglycerol molecule

Formation of a Triacylglycerol

A triacylglycerol actually is a triester of glycerol and fatty acids. Like all esters, it is formed by reacting an **alcohol** with a **carboxylic acid**.

alcohol + carboxylic acid = ester

(To recall this reaction, remember the word "ACE.")

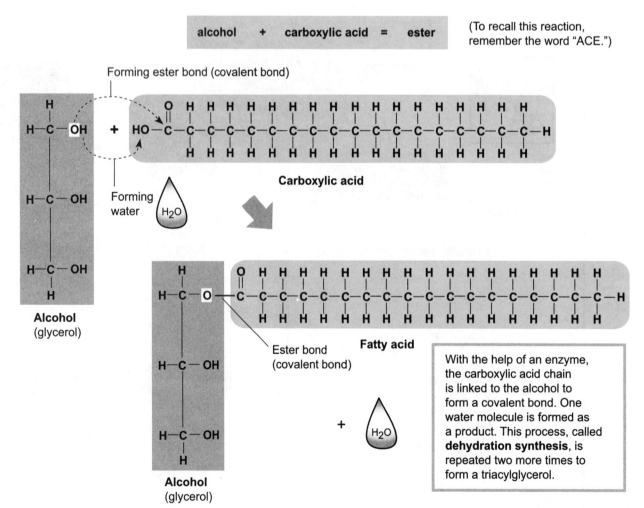

Forming ester bond (covalent bond)

Forming water H_2O

Carboxylic acid

Alcohol (glycerol)

Ester bond (covalent bond)

Fatty acid

Alcohol (glycerol)

+ H_2O

With the help of an enzyme, the carboxylic acid chain is linked to the alcohol to form a covalent bond. One water molecule is formed as a product. This process, called **dehydration synthesis**, is repeated two more times to form a triacylglycerol.

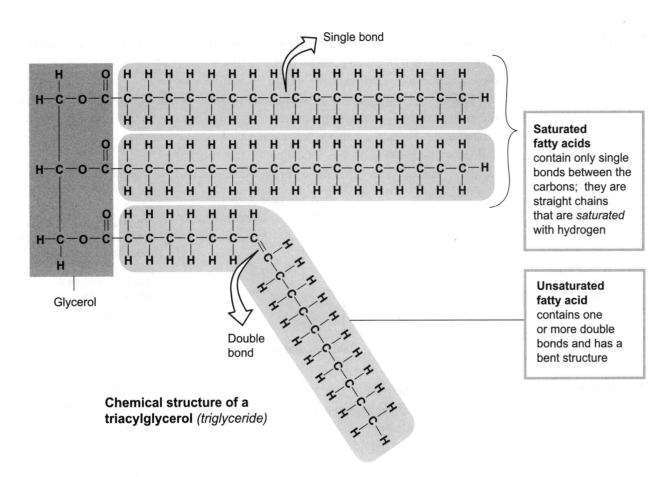

Single bond

Saturated
fatty acids
contain only single
bonds between the
carbons; they are
straight chains
that are *saturated*
with hydrogen

Glycerol

Double
bond

Unsaturated
fatty acid
contains one
or more double
bonds and has a
bent structure

**Chemical structure of a
triacylglycerol** *(triglyceride)*

Saturated and Unsaturated Fatty acids

In the illustration of the triacylglycerol above, the first two fatty acid chains are **saturated** and the third is **unsaturated**. Saturated fatty acids are so named because the carbon atoms are bonded to as many hydrogens as possible, hence, *saturated*. This makes the chains straight. Unsaturated fatty acids have one or more double bonds between carbon atoms. Because more hydrogens could be added, causing the double bond to break, fatty acids with double bonds are called *unsaturated*. The double bond gives the chain a bent structure. The term *polyunsaturated* refers to the presence of two or more double bonds.

On the grocery shelf it's easy to distinguish foods containing saturated from unsaturated fatty acids. Saturated fatty acids, like lard, are derived from animal fats and are solid at room temperature while unsaturated fatty acids, like corn oil, are derived from plants and are liquid at room temperature. As a general rule, a diet rich in unsaturated fatty acids is healthier because it may help prevent coronary artery disease.

Description

Steroids have a rigid ring structure and are the only class of lipids that do not contain fatty acid chains. Four fused carbon rings comprise the core structure of every steroid. Various side chains branching from these rings make one steroid chemically different from another.

In the body, many steroids act as chemical messengers called hormones. For example, the testes in the male produce the sex horomone testosterone, and the ovaries in the female produce estrogen and progesterone. The adrenal glands, found on top of the kidneys, produce the stress hormone cortisone and aldosterone—a sodium and potassium ion regulator.

In the clinical setting, patients are prescribed the steroid prednisone as an anti-inflammatory drug.

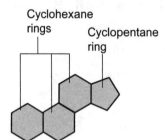

> The **basic steroid structure** contains three cyclohexane rings bonded to one cyclopentane ring. All steroids contain this basic ring structure.

Basic steroid structure

Cholesterol

Cholesterol is a steroid found in the animal products we consume and is made in small quantities by liver cells. It is a component in cell membranes that makes them more rigid and as a precursor to make other steroid compounds. As shown in the illustration below, cholesterol is a precursor in a multistep process to make the steroidal sex hormones progesterone, testosterone, and estrogen.

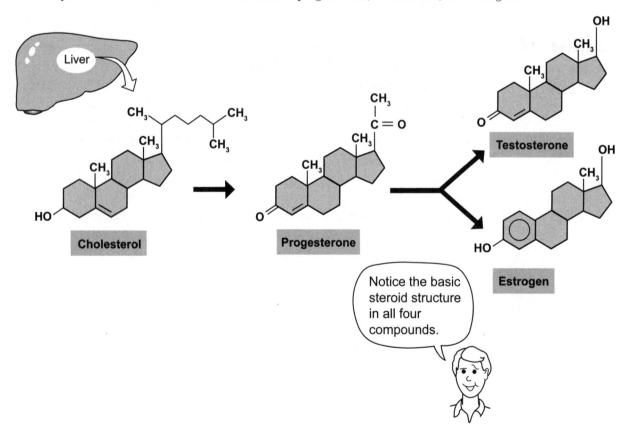

Sex Hormones: Testosterone and Estrogen

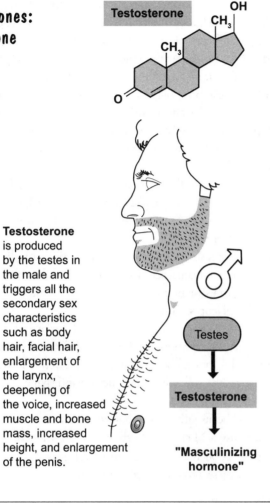

Testosterone is produced by the testes in the male and triggers all the secondary sex characteristics such as body hair, facial hair, enlargement of the larynx, deepening of the voice, increased muscle and bone mass, increased height, and enlargement of the penis.

Testes → Testosterone → "Masculinizing hormone"

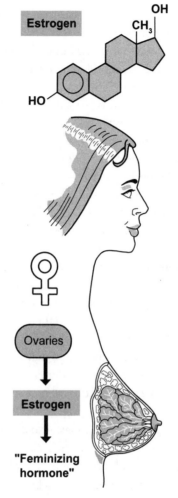

Ovaries → Estrogen → "Feminizing hormone"

Estrogen is produced by the ovaries in the female and triggers all the secondary sex characteristics such as body hair, pubic hair, development of the breasts, widening of the hips, increased fat deposits around the hips, buttocks, and thighs, increased height, and enlargement of the vulva.

Notice the similarities in the chemical structures of testosterone and estrogen.

Anabolic Steroids

Anabolic steroids are banned in the United States but are still used by some body builders and athletes to build muscle mass more quickly. Although they are effective, if used over the long term, their side effects include serious health hazards such as high blood pressure, decreased sperm production, and permanent liver damage.

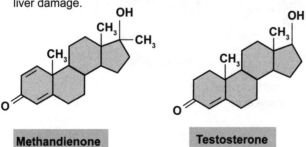

Methandienone   **Testosterone**

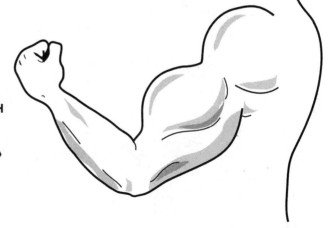

Methandienone is one example of an anabolic steroid. Like all anabolic steroids, it is a derivative of testosterone. Compare their similar chemical structures.

Description

Every cell in your body is enclosed by an envelope called a **cell membrane**. It is the gateway through which substances enter or exit any cell. A cell membrane is a double layer of lipids referred to as the **glycerophospholipid bilayer** or, more simply, the **lipid bilayer**. The **glycerophospholipid molecule** is the repeating unit within the cell membrane. Each molecule has a **polar head** and a **nonpolar tail** made of two fatty acid chains. Typically, one chain is saturated and straight and the other is unsaturated and bent. The polar head is hydrophilic ("water loving"), so it is attracted to water molecules, whereas the nonpolar tail is hydrophobic ("water fearing"), so it is not attracted to water molecules. As a result, the nonpolar tails position themselves away from water in the middle of the membrane while the polar heads position themselves toward polar water molecules on the outer and inner surfaces of the cell membrane. Embedded and scattered within the lipid bilayer is another type of lipid—**cholesterol molecules**—which provide some rigidity for the cell membrane.

The consistency of the cell membrane is referred to as a **fluid mosaic model**, as it is pliable. If you were microscopically small and could fall on its surface, it would feel more like a waterbed than a hard wooden floor. This is because the glycerophospholipid molecules have bent, unsaturated fatty acid chains, which prevent them from stacking together and allow for movement instead.

Proteins are another important part of the cell membrane. Those that span the entire lipid bilayer are called **integral proteins** and serve a variety of functions. Some act as channels through which only specific types of ions pass. Others act as receptors for hormones or neurotransmitters. Some integral proteins have carbohydrate chains attached to them and are called **glycoproteins**.

Peripheral proteins do not span the bilayer. Instead, they connect to only one surface of the membrane. Some act as enzymes, and others may serve as a structural component of the cytoskeleton. The **cytoskeleton** is composed of different types of protein filaments and is located beneath the lipid bilayer. It serves as a kind of scaffolding that supports the cell membrane and gives the cell a defined structure and shape.

Glycerophospholipid Molecule Structure

The glycerophospholipid molecule has a glycerol backbone with a phosphate group, and an alcohol attached to it. Branching from the glycerol backbone are two fatty acid chains. As shown below, the icon representing a glycerophospholipid molecule uses a sphere for the polar head, which consists of glycerol, a phosphate, and the alcohol, and the nonpolar tail is shown as two long strands—the fatty acid chains.

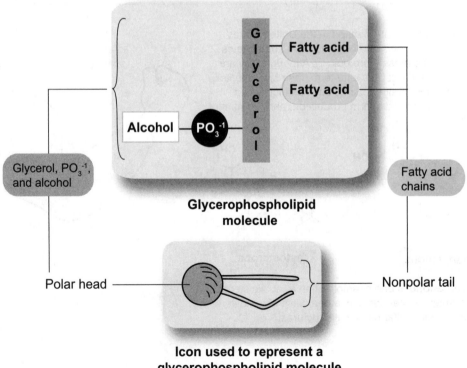

Glycerophospholipid molecule

Icon used to represent a glycerophospholipid molecule

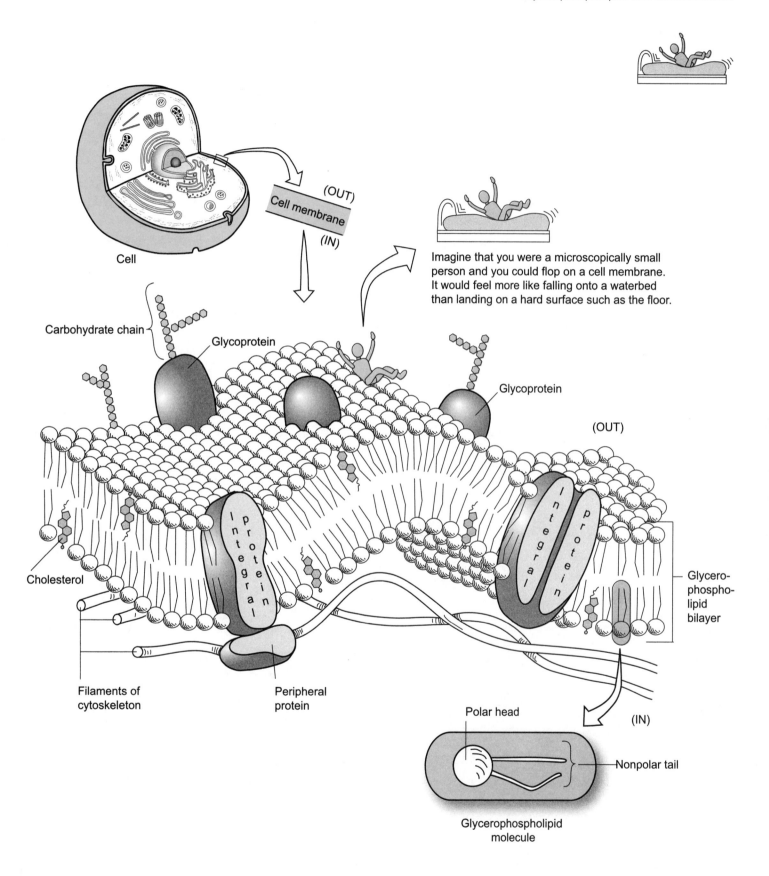

Cell

(OUT)
Cell membrane
(IN)

Imagine that you were a microscopically small person and you could flop on a cell membrane. It would feel more like falling onto a waterbed than landing on a hard surface such as the floor.

Carbohydrate chain

Glycoprotein

Glycoprotein

(OUT)

Cholesterol

Integral protein

Integral protein

Glycero-phospho-lipid bilayer

Filaments of cytoskeleton

Peripheral protein

Polar head

(IN)

Nonpolar tail

Glycerophospholipid molecule

Bioenergetics

Description

Metabolism is the sum total of all the chemical reactions that occur in the body. It consists of two different parts—catabolic reactions and anabolic reactions. **Catabolic reactions** break down biomolecules into simpler building blocks (e.g., triacylglycerols into fatty acids and glycerol), and **anabolic reactions** use these building blocks to synthesize more complex biomolecules (e.g., amino acids into proteins). Energy metabolism focuses on the elements of metabolism that generate ATP. As a general overview of energy metabolism, let's examine its three main phases:

PHASE 1:
The Release of Nutrients in the Digestive Tract Lumen

The digestion of three major nutrients—triacylglycerols, complex carbohydrates, and proteins—is shown below. As these macromolecules are ingested, they enter the digestive tract and gradually are broken down into their fundamental building blocks:

- Proteins $\rightarrow \rightarrow \rightarrow \rightarrow \rightarrow \rightarrow$ amino acids
- Complex carbohydrates $\rightarrow \rightarrow$ glucose (and other simple sugars)
- Triacylglycerols $\rightarrow \rightarrow \rightarrow \rightarrow$ monoglycerides and fatty acids

This phase of catabolism is catalyzed by digestive enzymes in the digestive tract. These building blocks now are small enough to be used by cells. After being absorbed into the bloodstream through microscopic structures called **villi** on the inner lining of the small intestine, they are transported to various cells throughout the body.

PHASE 2:
The Fate of Nutrients Within Cells of Various Tissues

Glucose has three possible fates:
- **Catabolized:** When the body requires energy, glucose is catabolized to produce ATP molecules. The three stages of glucose catabolism are: glycolysis, the citric acid cycle, and the electron transport system.
- **Stored:** This takes place in the form of **glycogen**, a long chain of glucose molecules. Skeletal muscle cells and liver cells store lots of glycogen when excess glucose is present in the blood.
- **Converted:** Glucose can be converted into several **amino acids**, which may be used to make proteins. Alternatively, when storage of glycogen is maximized, liver cells can convert glucose into **glycerol** and **fatty acids** to form **triglycerides**. The triglycerides then are deposited in fatty tissue.

Fatty acids and **glycerol** have one of two general fates:
- **Converted/catabolized:**

　　Glycerol can be converted into glyceraldehyde–3–phosphate, one of the compounds formed during the process of glycolysis. Then, depending on the ATP levels in the cell, it is either converted into **pyruvate** (via glycolysis) or **glucose** (via gluconeogenesis).

　　Fatty acids follow a different pathway than glycerol. They can be converted into **acetyl CoA**, which then enters the citric acid cycle to produce ATP.
- **Stored:** glycerol and fatty acids recombine to form triacylglycerols within liver cells and fat cells.

Amino acids have three different fates:
- **Anabolized:** Amino acids may be used to produce **proteins**.
- **Converted:** Some amino acids can be converted into **acetyl CoA** and enter the citric acid cycle to produce ATP. When there are excess amino acids, they can be converted into either **glucose** (via gluconeogenesis) or **fatty acids**.
- **Deaminated:** Liver cells can remove the amino group from some amino acids to produce carboxylic acids and ammonia (NH_3) in a process called deamination. Then the carboxylic acids can enter different stages of the citric acid cycle.

PHASE 3:
Aerobic Catabolism of Nutrients in the Mitochondria of Cells

Aerobic respiration refers to the two parts of cellular respiration that occur inside the mitochondrion: the citric acid cycle (see p. 228) and the electron transport system (see p. 230). **Acetyl CoA**, a key chemical intermediate in metabolism, is generated from fatty acids, pyruvate, and some amino acids within the mitochondrion. Once acetyl CoA is produced, it may enter the citric acid cycle as part of normal catabolism.

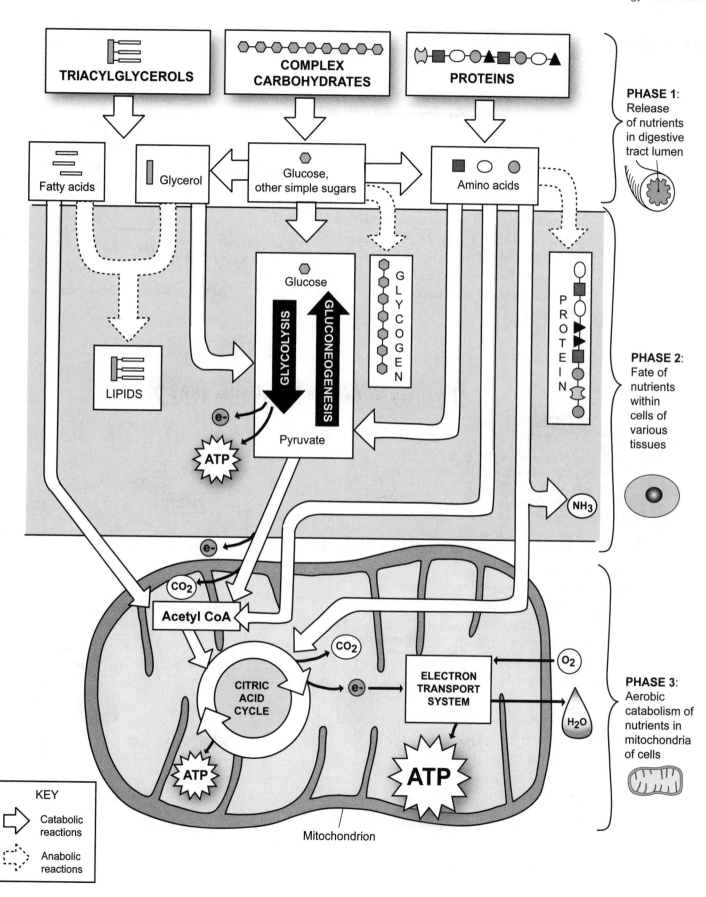

Description

We will examine three important coenzymes in metabolic pathways: NAD⁺, FAD, and coenzyme A. They have the following features in common:

Each:

- contains **ADP**
- is a B–vitamin derivative
- is involved in the metabolism of food components such as carbohydrates, proteins, and fatty acids to yield energy.

Vitamins are small organic compounds that we are unable to make and therefore must consume. Several coenzymes involved in glycolysis, the citric acid cycle, and the electron transport system are composed of B vitamins.

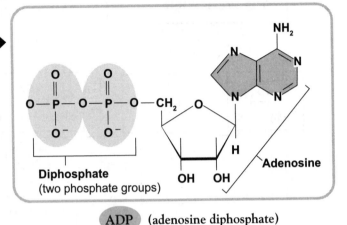

Diphosphate (two phosphate groups)   **Adenosine**

ADP (adenosine diphosphate)

Nicotinamide Adenine Dinucleotide (NAD⁺)

Analogy:

- NAD⁺ is like a car with a trailer because it picks up its cargo (2 electrons and 1 proton) and delivers it to another compound.

Unique function:

- NAD⁺ accepts 2 $e-$ and 1 H⁺ from a compound to form NADH. (A second H⁺ is lost to the surrounding solution)

$$NAD^+ + 2\,H^+ + 2\,e- \longrightarrow NADH + H^+$$

lost to surrounding solution

Derived from:

- Niacin (a B–vitamin)

$2\,H^+ + 2\,e-$

Nicotinamide (niacin)

Structure:

NAD⁺ has 3 parts:
- Nicotinamide
- Ribose
- ADP

ADP — CH₂

Ribose

OH   OH

lost to surrounding solution

OH   OH

218

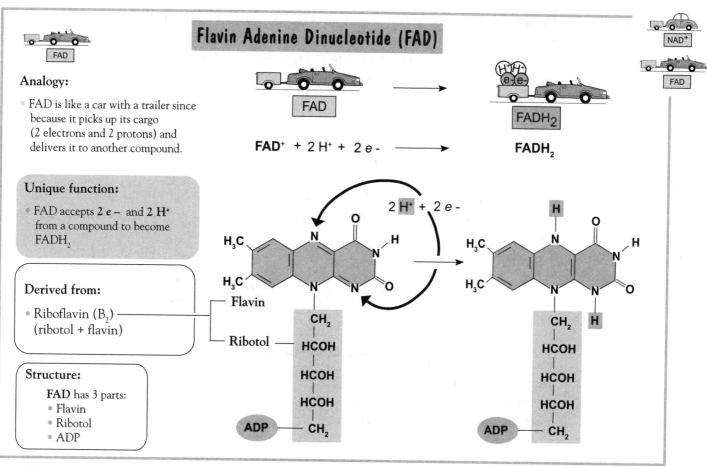

Flavin Adenine Dinucleotide (FAD)

Analogy:

- FAD is like a car with a trailer since because it picks up its cargo (2 electrons and 2 protons) and delivers it to another compound.

FAD $+ 2H^+ + 2e-$ ⟶ **FADH$_2$**

Unique function:

- FAD accepts **2 $e-$** and **2 H$^+$** from a compound to become FADH$_2$

Derived from:

- Riboflavin (B$_2$) (ribotol + flavin)

Flavin

Ribotol

Structure:

FAD has 3 parts:
- Flavin
- Ribotol
- ADP

$2 H^+ + 2 e-$

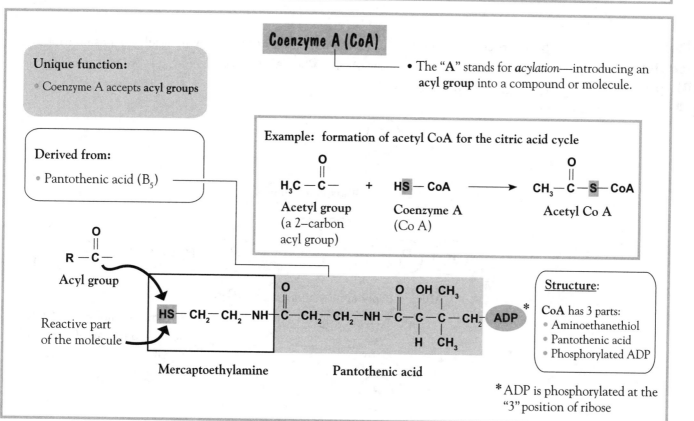

Coenzyme A (CoA)

- The "**A**" stands for *acylation*—introducing an **acyl group** into a compound or molecule.

Unique function:

- Coenzyme A accepts **acyl groups**

Derived from:

- Pantothenic acid (B$_5$)

Example: formation of acetyl CoA for the citric acid cycle

Acetyl group (a 2–carbon acyl group) + Coenzyme A (Co A) ⟶ Acetyl Co A

Acyl group

Reactive part of the molecule

Mercaptoethylamine

Pantothenic acid

Structure:

CoA has 3 parts:
- Aminoethanethiol
- Pantothenic acid
- Phosphorylated ADP

*ADP is phosphorylated at the "3" position of ribose

Description

Phosphorylation is a common process in metabolic pathways involving the addition of an inorganic phosphate group (P_i) to another compound. The opposite process, **dephosphorylation**, occurs when an inorganic phosphate group is removed from a compound. These two processes are often linked in an ongoing cycle. Phosphorylation is commonly used to regulate the activity of enzymes to either activate or inactivate them.

Concept and Example

Many phosphorylation reactions are catalyzed by enzymes called **kinases**.

In the liver, the enzyme **pyruvate kinase** becomes inactivated after it is phosphorylated.

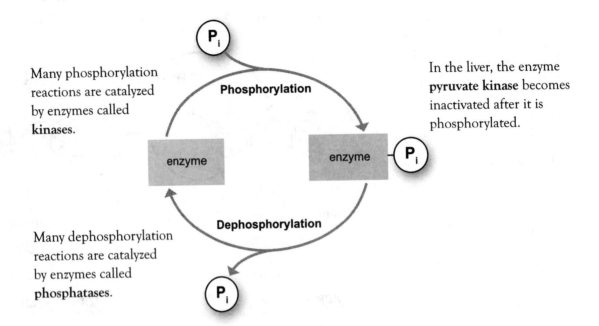

Many dephosphorylation reactions are catalyzed by enzymes called **phosphatases**.

Other Examples: ADP and ATP

Phosphorylation also occurs during the formation of the energy molecule adenosine triphosphate (ATP) from adenosine diphosphate (ADP). When ATP is used as a fuel, it is hydrolyzed and undergoes dephosphorylation to become ADP.

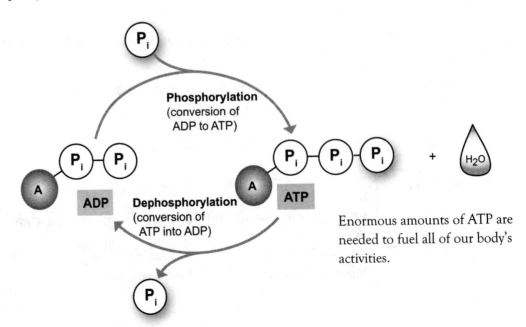

Enormous amounts of ATP are needed to fuel all of our body's activities.

Oxidative Phosphorylation:

- The process in which large amounts of ATP are made from ADP in mitochondria during the last phase of glucose catabolism, called the electron transport system.
- The term "oxidative" refers to the oxidation of reduced coenzymes (NADH and $FADH_2$), which deliver electrons to this process.

Roadmap

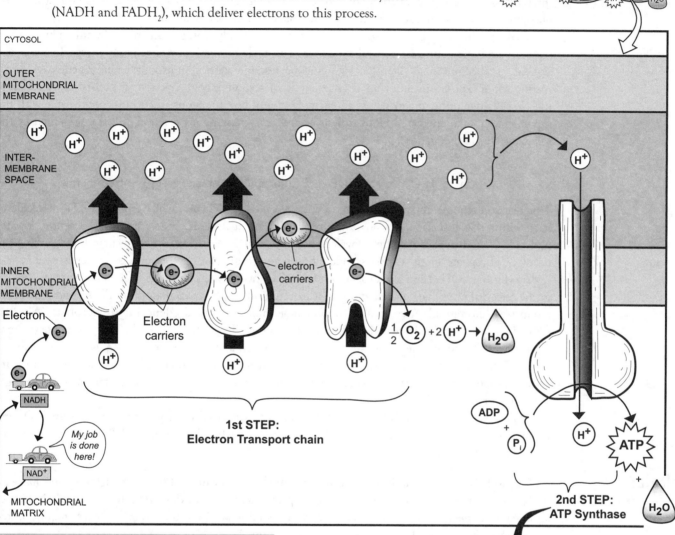

1st STEP: ELECTRON TRANSPORT CHAIN

- Electrons (e-) from NADH and $FADH_2$ are delivered to membrane proteins in the inner mitochondrial membrane, which serve as electron carriers.
- The energy from this electron transfer is used to force H^+ out of the matrix and into the intermembrane space. The resulting proton gradient serves as a source of potential energy.
- The final electron acceptor, oxygen (O_2), reacts with H^+ to form water (H_2O).

2nd STEP: ATP synthase

- The H^+ concentration gradient causes the H^+ to diffuse back through the inner mitochondrial membrane, but the only passageway to the matrix is through the ATP synthase enzyme.
- This flow of H^+ through ATP synthase, like water through a dam, drives the phosphorylation of ADP to form ATP.

ATP Synthase: DAM ANALOGY

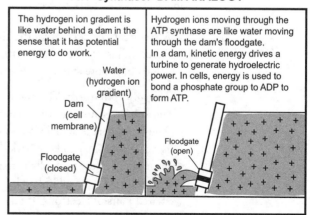

The hydrogen ion gradient is like water behind a dam in the sense that it has potential energy to do work.

Hydrogen ions moving through the ATP synthase are like water moving through the dam's floodgate.
In a dam, kinetic energy drives a turbine to generate hydroelectric power. In cells, energy is used to bond a phosphate group to ADP to form ATP.

Water (hydrogen ion gradient)

Dam (cell membrane)

Floodgate (closed)

Floodgate (open)

Description

Adenosine triphosphate (ATP) is the universal energy currency for all cells. Look at the illustration of ATP on the facing page to understand its structure. It is classified as a nucleotide and consists of three parts: (1) **adenine base**, (2) **ribose sugar**, and (3) **phosphate groups**—three total. The term "adenosine" refers to the adenine base and ribose sugar bonded together, and the term "triphosphate" comes from the three phosphate groups covalently bonded to each other. There is a net negative charge on each of the phosphate groups, so they repel each other and make ATP relatively unstable.

Cells are highly organized structures that are constantly doing work to maintain their physical structure and carry out their general functions. This cycle of regular work requires energy. In biological systems, it is common to couple a spontaneous reaction with a nonspontaneous reaction. A metal rusting when exposed to moist air is an example of a spontaneous reaction (it occurs all by itself). A muscle cell attempting to contract is an example of a nonspontaneous reaction (it occurs only with the input of additional energy). ATP hydrolysis occurs as a spontaneous reaction and is represented by the following chemical equation:

$$\text{ATP} + \text{H}_2\text{O} \xrightarrow{\text{ATPase}} \text{ADP} + \text{P}_i \text{ (inorganic phosphate group)} + \textbf{free energy}$$

This reaction requires the assistance of an enzyme called an **ATPase**, which binds an ATP molecule to itself with a shape-specific fit like a lock and key. In any hydrolysis ("water splitting") reaction, a water molecule is used to cleave a single covalent bond. In this case, it cleaves the covalent bond linking the terminal phosphate group to the second phosphate group. During this process, water is split into a **hydroxyl group** (OH^-) and a **hydrogen ion** (H^+). The hydroxyl group binds to the phosphorus atom (P) in the terminal phosphate group, and the hydrogen binds to the oxygen atom on the second phosphate group (see illustration). As a result, a more stable molecule called **adenosine diphosphate (ADP)** is formed as a result of the decreased repulsion between the phosphate groups. The free phosphate group is often transferred to another substrate or to an enzyme. In addition, some free energy is released in the process. The free energy is used to drive processes that must occur in cells, such as muscle contraction, active transport of substances across cell membranes, movement of a sperm cell's tail, manufacturing a hormone, or anything else a cell has to do.

ATP is manufactured inside your cells via the energy derived from foods that are ingested. During this process, called cellular respiration, ATP is formed by bonding ADP and P_i through a series of **oxidation–reduction** reactions.

Analogies

ATP hydrolysis is also like an **investment in a profitable stock or mutual fund**. Though you must make an initial investment of your own money, you will reap a greater reward (dividend) in the end. Similarly, it takes an initial investment of energy to break the covalent bond in ATP, but the result is a larger amount of free energy when the new bonds in the final products are formed.

Study Tips

Unfortunately, the function of ATP and the process of ATP hydrolysis often are explained incorrectly in many textbooks. Let's correct some common misconceptions:

Misconception #1: **Breaking chemical bonds releases energy.**
Actually, just the opposite is true. During a chemical reaction, the formation of new chemical bonds in a more stable product results in the release of energy.

Misconception #2: **ATP has a special "high energy phosphate bond" between the second phosphate group and the terminal phosphate group.**
The bond here is actually a covalent bond. This statement gives the false impression that this bond is ready to fly apart like a "jack-in-the-box." In fact, bonds are a force that hold atoms together so an input of energy actually is required to break a chemical bond. Instead, the three phosphate groups in ATP are all linked by relatively strong bonds called covalent bonds, which represent shared pairs of electrons. It takes energy to break these bonds.

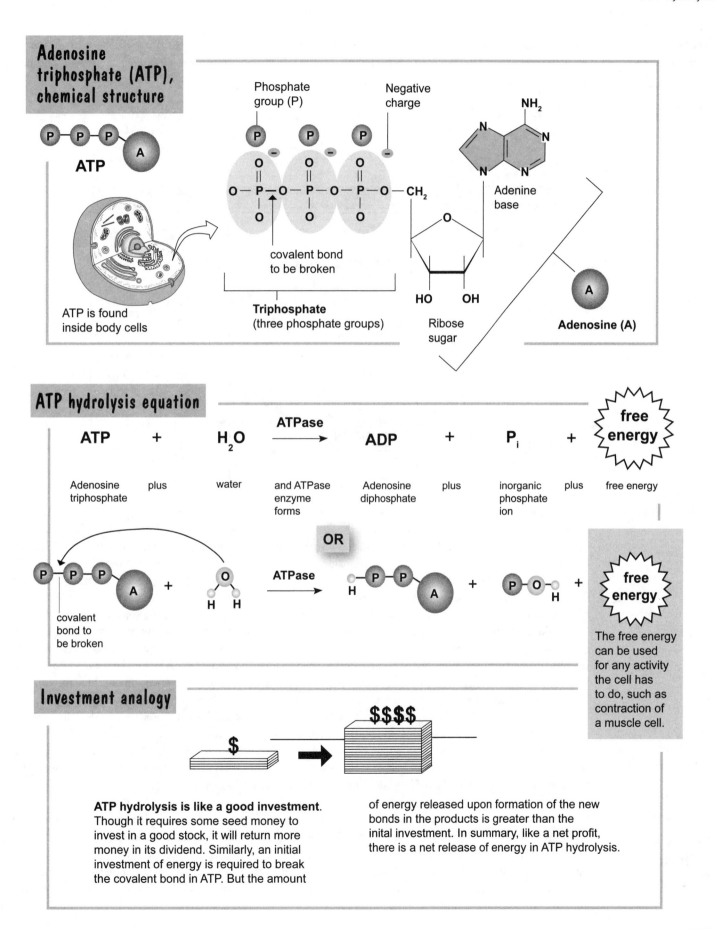

Adenosine triphosphate (ATP), chemical structure

ATP

ATP is found inside body cells

Phosphate group (P)

Negative charge

NH_2

Adenine base

covalent bond to be broken

Triphosphate (three phosphate groups)

Ribose sugar

Adenosine (A)

ATP hydrolysis equation

$$ATP + H_2O \xrightarrow{ATPase} ADP + P_i + \text{free energy}$$

| Adenosine triphosphate | plus | water | and ATPase enzyme forms | Adenosine diphosphate | plus | inorganic phosphate ion | plus | free energy |

OR

$$\xrightarrow{ATPase}$$

covalent bond to be broken

free energy

The free energy can be used for any activity the cell has to do, such as contraction of a muscle cell.

Investment analogy

$ $$$

ATP hydrolysis is like a good investment. Though it requires some seed money to invest in a good stock, it will return more money in its dividend. Similarly, an initial investment of energy is required to break the covalent bond in ATP. But the amount of energy released upon formation of the new bonds in the products is greater than the inital investment. In summary, like a net profit, there is a net release of energy in ATP hydrolysis.

Overall Equation for Oxidation of Glucose

In the process of cellular respiration, 1 molecule of glucose reacts with 6 oxygen molecules through a series of chemical reactions to produce 6 carbon dioxide molecules, 6 water molecules, 36 molecules of ATP, and heat energy, which is lost to the environment.

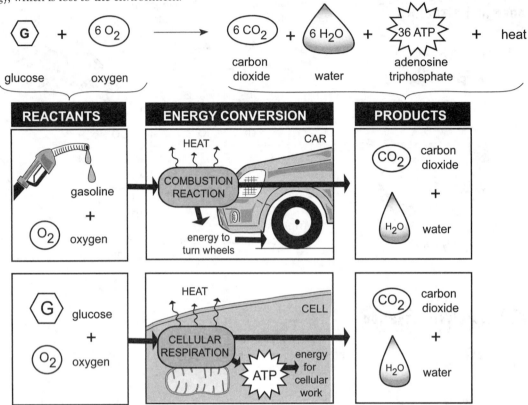

$$\boxed{G} + \boxed{6\ O_2} \longrightarrow \boxed{6\ CO_2} + \boxed{6\ H_2O} + \boxed{36\ ATP} + heat$$

glucose   oxygen   carbon dioxide   water   adenosine triphosphate

REACTANTS — **ENERGY CONVERSION** — **PRODUCTS**

gasoline + oxygen → COMBUSTION REACTION (CAR) / HEAT / energy to turn wheels → carbon dioxide + water

glucose + oxygen → CELLULAR RESPIRATION (CELL) / HEAT / ATP / energy for cellular work → carbon dioxide + water

Description

The process of cellular respiration is like the combustion of gasoline in a car engine. Both require a fuel that must be chemically reacted with oxygen. Glucose is the fuel for the cell, whereas gasoline is the fuel for the car engine. The car engine uses a spark plug to ignite the combustible oxygen/gasoline mixture, causing it to produce a small explosion, which moves a piston up and down inside a metal cylinder. This kinetic energy is transferred to an axle, causing it to spin. Because the wheels are attached to the axle, they spin as the axle spins. The direct products of this reaction—carbon dioxide and water vapor—are released in the exhaust gases while heat is lost to the external environment. In summary, the potential energy in gasoline is converted into kinetic energy, which propels the car.

Cellular respiration works in a similar manner. Instead of having all the energy released at once, glucose is gradually broken down through a long series of chemical reactions. The overall purpose is to trap as much energy as possible from glucose in the form of an energy molecule called ATP. Then the ATP can be used to do cellular work such as contracting a muscle cell. During the process, some heat is lost to the environment. Like the combustion of gasoline, the final products of this chemical process are carbon dioxide and water. In summary, some of the chemical bond energy in glucose is transferred slowly into ATP molecules, which can be used to do cellular work.

Cellular respiration is a series of oxidation–reduction reactions, or redox reactions (see p. 78). These involve the transfer of electrons from one substance to another. In this case, glucose is oxidized because it loses electrons to oxygen. Oxygen is reduced because it gains the electrons lost from glucose. To distinguish oxidation from reduction remember the words OIL RIG = Oxidation Is Loss, Reduction Is Gain.

Efficiency Rating of Cellular Respiration

Cellular respiration incorporates three processes: (1) glycolysis, (2) citric acid cycle, and (3) the electron transport system (E.T.S.). Each of these processes will be discussed in more detail in separate modules.

The process of cellular respiration is more efficient than a car engine. During the oxidation of glucose, 686 kilocalories of energy are released. Of this total, 278 are captured in the bonds of ATP molecules, giving it an efficiency rating of 41%. The remaining 59% is lost as heat. Compared to a typical car engine that is between 10–30% efficient, this rating is looking pretty good!

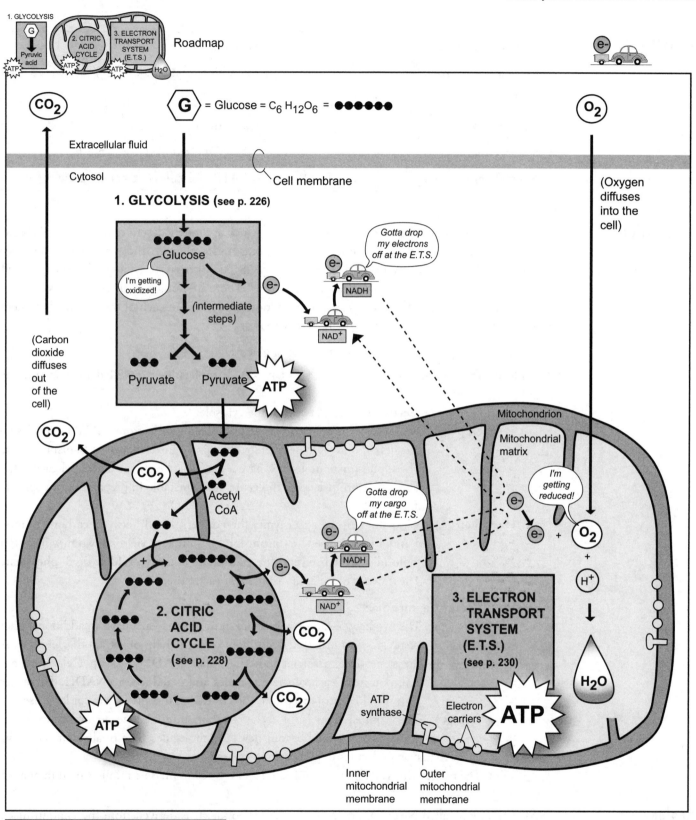

Roadmap

G = Glucose = $C_6H_{12}O_6$ = ●●●●●●

1. GLYCOLYSIS (see p. 226)

Glucose

I'm getting oxidized!

(intermediate steps)

Pyruvate   Pyruvate

Gotta drop my electrons off at the E.T.S.

ATP

Extracellular fluid

Cytosol

Cell membrane

(Oxygen diffuses into the cell)

(Carbon dioxide diffuses out of the cell)

Acetyl CoA

2. CITRIC ACID CYCLE (see p. 228)

Gotta drop my cargo off at the E.T.S.

Mitochondrion

Mitochondrial matrix

I'm getting reduced!

3. ELECTRON TRANSPORT SYSTEM (E.T.S.) (see p. 230)

ATP synthase

Electron carriers

ATP

Inner mitochondrial membrane

Outer mitochondrial membrane

KEY

e− = electron

H^+ = hydrogen ion

NAD+ = nicotinamide adenine dinucleotide (coenzyme)

FUN FACT: The carbon atom in each molecule of carbon dioxide gas that you exhale was originally one of the carbon atoms in glucose.

Overview

Location within cell:	cytosol
Aerobic or anaerobic?	anaerobic (doesn't require O_2)
Initial reactant:	glucose
Final product(s):	• 2 pyruvates
Side products:	• 2 NADH (reduced coenzymes)
Net yield of energy:	• 2 ATP molecules (4 created – 2 used)

Description

The term **glycolysis** means the "splitting of glucose," which accurately describes the process. It begins with one glucose molecule containing 6 carbon atoms and ends with the formation of two new 3-carbon molecules called **pyruvate**. This process is catalyzed by various enzymes in the **cytosol** of cells and yields a net gain of 2 ATP molecules and 2 NADH. There are numerous intermediate steps.

Instead of presenting the various chemical reactions that occur in each of the intermediate steps, the process will be presented conceptually in two phases:

Phase 1: Energy investment phase (splitting of glucose)

Key idea: **Free energy from the hydrolysis of two ATP molecules is invested to help split the original glucose molecule.**

- Glucose enters the cell and remains in the **cytosol**.
- Step ① With the help of an enzyme, 2 phosphate groups from 2 ATP molecules are transferred onto glucose, thereby changing it into **fructose–1,6–bisphosphate**. These phosphate molecules have a net negative charge and repel each other. The repulsion makes the molecule more unstable in preparation for splitting it.
- Step ② With the help of an enzyme, the covalent bond between carbons number 3 and 4 in fructose-1,6-bisphosphate is broken, yielding two new 3-carbon molecules: **dihydroxyacetone phosphate** and **glyceraldehyde 3-phosphate**. These two molecules are isomers of each other.

Phase 2: Energy Capture Stage

- Step ③ The two new 3-carbon molecules from phase 1 each have an additional phosphate group transferred onto them with the help of yet another enzyme. At the same time, a carrier molecule called NAD^+ picks up 2 electrons and 1 proton from each of the molecules and is reduced to **NADH**. These electrons can be transported into the electron transport system for later use.
- Step ④ The 2 phosphates from each 3-carbon intermediate are transferred onto ADP to form a total of four new ATP molecules. The final result is that two new molecules of pyruvate are created.
- The net gain in ATP for glycolysis is 2 ATP (**4 created** in phase 2 minus **2 used** in phase 1)

Analogy

NAD^+ is an ion called Nicotinamide Adenine Dinucleotide. It is derived from the vitamin niacin and functions as a carrier ion that transports electrons to the electron transport system (E.T.S.). When NAD^+ is reduced (gains electrons), it becomes **NADH**. The NAD^+ ion is like a **car pulling a trailer with no cargo** while **NADH** is like a **car filled with its cargo of two electrons (and one proton)** (see p. 218).

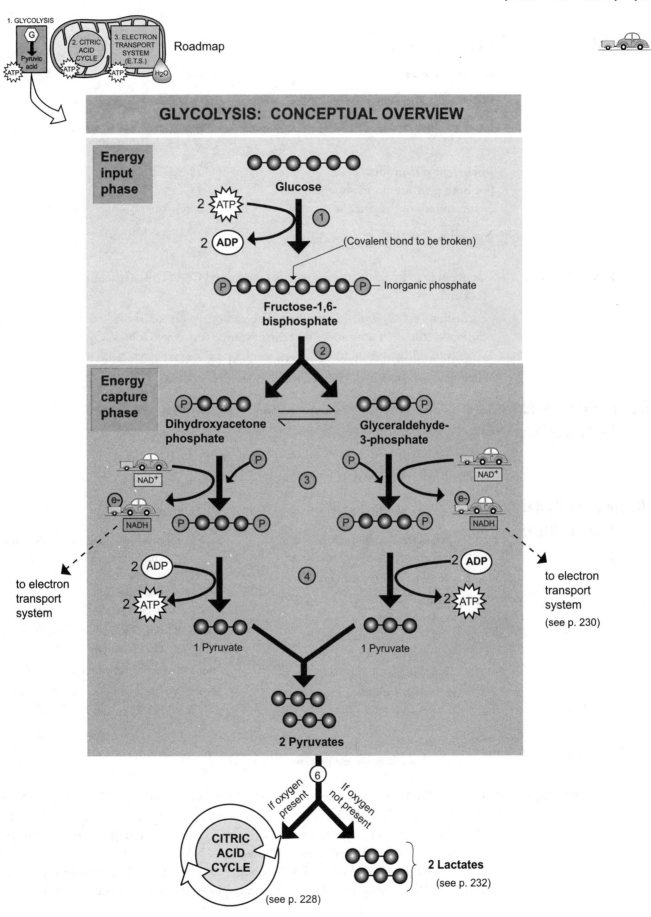

GLYCOLYSIS: CONCEPTUAL OVERVIEW

Energy input phase

Glucose

2 ATP

2 ADP

(Covalent bond to be broken)

P — Inorganic phosphate

Fructose-1,6-bisphosphate

Energy capture phase

Dihydroxyacetone phosphate

Glyceraldehyde-3-phosphate

NAD⁺

NADH

to electron transport system

NAD⁺

NADH

to electron transport system (see p. 230)

2 ADP

2 ATP

2 ADP

2 ATP

1 Pyruvate

1 Pyruvate

2 Pyruvates

If oxygen present

If oxygen not present

CITRIC ACID CYCLE

(see p. 228)

2 Lactates
(see p. 232)

Roadmap

1. GLYCOLYSIS

G

Pyruvic acid

ATP

2. CITRIC ACID CYCLE

3. ELECTRON TRANSPORT SYSTEM (E.T.S.)

ATP

ATP

H₂O

Overview

Location within cell:	mitochondrion
Aerobic or anaerobic?	aerobic (indirectly)
Initial reactants/substrates:	oxaloacetate and acetyl CoA
Final product:	oxaloacetate (which condenses with acetyl CoA to form citrate)
Important side products: (for **both** pyruvates from glycolysis or two rounds of the citric acid cycle)	• 6 NADH, 2 $FADH_2$ • 6 CO_2 molecules (4 during the cycle; 2 in formation of acetyl CoA)
Net yield of energy:	• 2 ATP molecules

3 Key Events:

1 Conversion of pyruvate (from glycolysis) into **acetyl CoA**. This is what links glycolysis to the citric acid cycle.

2 Formation of CO_2 as a side product. The carbon in carbon dioxide can be traced back to the carbon in glucose. Think about that the next time you exhale some CO_2!

3 Formation of many **reduced coenzymes (NADH, $FADH_2$)**. These reduced coenzymes link the citric acid cycle to the electron transport system.

Before Citric Acid Cycle Begins:

Pyruvate is first converted into **acetic acid** and then into **acetyl coenzyme A** (CoA).

—CO_2 is released in the process.

—CoA is used to carry the 2-carbon remnant of pyruvate oxidation.

—NAD^+ is reduced to **NADH** by picking up 2 electrons.

Citric Acid Cycle: Step by Step:

NOTE: Each of the steps in the citric acid cycle requires the help of a different enzyme.

Step 1: **Oxaloacetate** condenses with **acetyl coenzyme A** (CoA) to form **citrate**.

Step 2: **Citrate** is converted into **isocitrate**.

Step 3: **Isocitrate** is converted into **alpha (α)-ketoglutarate**.
- NAD^+ is reduced to NADH
- CO_2 is released in the process

Step 4: **Alpha (α)-ketoglutarate** is converted into **succinyl CoA**.
- CoA enters this step to carry the succinyl group
- NAD^+ is reduced to NADH
- CO_2 is released in the process

Step 5: **Succinyl CoA** is converted into **succinate**.
- Phosphate group is transferred to GDP to create GTP
- Hydrolysis of GTP transfers phosphate group to ADP to form ATP

Step 6: **Succinate** is converted into **fumarate**.
- FAD is reduced to $FADH_2$

Step 7: **Fumarate** is converted into **malate**.

Step 8: **Malate** is converted into **oxaloacetate**.
- NAD^+ is reduced to NADH

Analogy:

NAD^+ and **FAD** are compounds that act like **shuttle buses**. NAD^+ is called Nicotinamide Adenine Dinucleotide, while **FAD** is called Flavin Adenine Dinucleotide. FAD is derived from vitamin B_2 (riboflavin). Both are classified as carrier molecules that are represented as different types of vehicles carrying a cargo. By picking up two electrons, NAD^+ gets reduced to **NADH** and **FAD** gets reduced to $FADH_2$. The NAD^+ or FAD molecule is like a **car pulling a trailer with no cargo**, and NADH or $FADH_2$ is like a **car filled with its cargo of electrons (and protons)** (see p. 219).

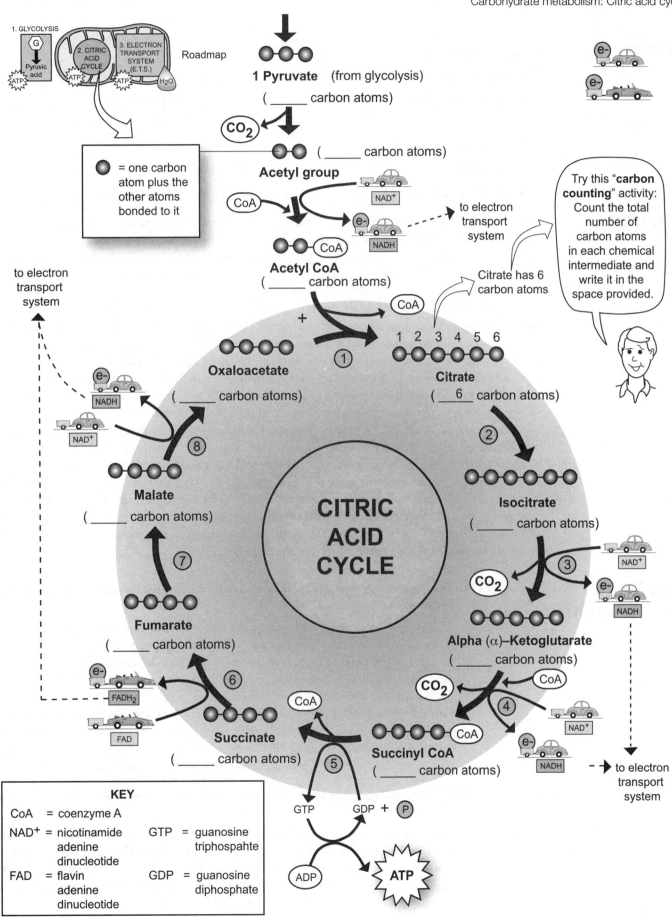

1. GLYCOLYSIS

2. CITRIC ACID CYCLE

3. ELECTRON TRANSPORT SYSTEM (E.T.S.)

Roadmap

= one carbon atom plus the other atoms bonded to it

1 Pyruvate (from glycolysis)
(_____ carbon atoms)

CO_2

Acetyl group
(_____ carbon atoms)

CoA

Acetyl CoA
(_____ carbon atoms)

to electron transport system

Try this "**carbon counting**" activity: Count the total number of carbon atoms in each chemical intermediate and write it in the space provided.

Citrate has 6 carbon atoms

to electron transport system

CITRIC ACID CYCLE

Oxaloacetate
(_____ carbon atoms)

1 2 3 4 5 6
Citrate
(__6__ carbon atoms)

Malate
(_____ carbon atoms)

Isocitrate
(_____ carbon atoms)

CO_2

Fumarate
(_____ carbon atoms)

Alpha (α)–Ketoglutarate
(_____ carbon atoms)

CO_2

CoA

Succinate
(_____ carbon atoms)

Succinyl CoA
(_____ carbon atoms)

to electron transport system

GTP   GDP + P

ADP   **ATP**

KEY

CoA = coenzyme A

NAD+ = nicotinamide adenine dinucleotide

FAD = flavin adenine dinucleotide

GTP = guanosine triphosphate

GDP = guanosine diphosphate

Overview

Location within cell:	mitochondrion
Aerobic or anaerobic?	aerobic (O_2 used **directly**)
Important side products:	• 10 NADH and 2 $FADH_2$ (total of 12 reduced coenzymes)
Final electron acceptor:	O_2 (oxygen)
Final product:	H_2O (water)
Net yield of energy:	**32 ATP** molecules (maximum)

Description

The **Electron Transport System (E.T.S.)** is the last step in cellular respiration yet it produces far more ATP—32—than either of the first two steps. E.T.S. accomplishes three major things: (1) uses the energy from stored electrons (e-) to create a potential energy gradient in the form of hydrogen ions (protons), (2) converts this potential energy into kinetic energy to produce lots of ATP, and (3) regenerates NAD^+ and FAD to allow glycolysis and the citric acid cycle to continue.

The reduced coenzymes (**NADH** and **$FADH_2$**) directly link the citric acid cycle to the E.T.S. They are carrying electrons from the original glucose molecule that started the whole process. These electrons are delivered to membrane proteins and other electron carriers in the inner mitochondrial membrane where the E.T.S. process occurs. The electrons are shuttled through a series of electron carriers, where they go from a high-energy state to a low-energy state. Oxygen serves as the final electron acceptor. As oxygen bonds with the electron, it also bonds to hydrogen ions (H^+) to create a water molecule in the matrix. Where do the hydrogen ions come from? They are in the solution of the matrix.

The electron carriers use the energy from the shuttled electrons to pump hydrogen ions from the mitochondrial matrix into the intermembrane space. The result is that a gradient of hydrogen ions is created, which serves as a source of potential energy. The primary way for the hydrogen ions to move quickly out of the intermembrane space is to flow through a pore in a membrane protein called an **ATP synthase**. As the hydrogen ions diffuse down their gradient by flowing through this pore, the potential energy is converted into kinetic energy. This kinetic energy is used to covalently bond a phosphate group (P_i) to an ADP molecule to produce an ATP molecule. Note that both the phosphate group and the ADP molecule are present already in the solution of the matrix.

2 Key Concepts and Their Analogies:

- **Concept #1:** **The high-energy electron provides energy to the electron carriers to create a hydrogen ion gradient.**

- **Analogy #1:** The electron being transported between the proteins in the inner mitochondrial membrane is like a **hot baked potato**. The hot potato has **thermal energy** (heat) just like the electron has **energy**. The **electron carriers in the inner mitochondrial membrane** are like a **row of people** standing next to each other. Imagine that the first person tosses the baked potato to the next in line, who then tosses it to the next, and so on. The result is that each person's hands absorb a little bit of the heat from the potato. Similarly, the electron transfers some energy to the electron carriers, which then is used to transport a hydrogen ion (or proton) from the mitochondrial matrix to the inner membrane space. The net result is that a hydrogen ion gradient is formed across the inner mitochondrial membrane.

- **Concept #2:** **The potential energy in the hydrogen ion gradient is converted into kinetic energy and used to create ATP.**

- **Analogy #2:** The **potential energy** in the hydrogen ion gradient is like **water behind a dam**. The **dam** is the **inner mitochondrial membrane**, and the **water behind the dam** is like the **gradient of hydrogen ions** in the intermembrane space. The **floodgate** in the dam is like the **ATP synthase** and the kinetic energy from the flow of water through the dam is like the **flow of hydrogen ions** through the pore in the ATP synthase protein.

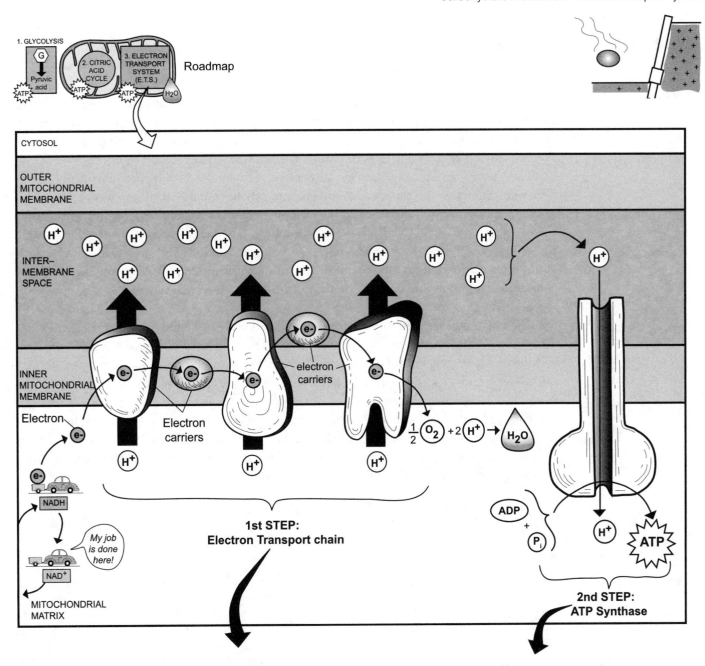

Roadmap

1. GLYCOLYSIS

G

Pyruvic acid

ATP

2. CITRIC ACID CYCLE

ATP

3. ELECTRON TRANSPORT SYSTEM (E.T.S.)

ATP

H_2O

CYTOSOL

OUTER MITOCHONDRIAL MEMBRANE

INTER– MEMBRANE SPACE

INNER MITOCHONDRIAL MEMBRANE

electron carriers

Electron

Electron carriers

$\frac{1}{2}$ O_2 + 2 H^+ → H_2O

$e-$

NADH

My job is done here!

NAD$^+$

MITOCHONDRIAL MATRIX

1st STEP:
Electron Transport chain

ADP + P_i

H^+

ATP

2nd STEP:
ATP Synthase

Electron Transport chain: HOT POTATO ANALOGY

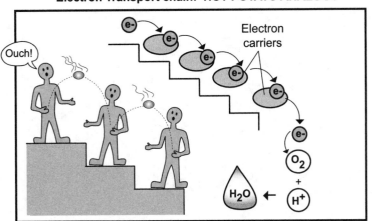

Ouch!

Electron carriers

O_2 + H^+

H_2O

ATP Synthase: DAM ANALOGY

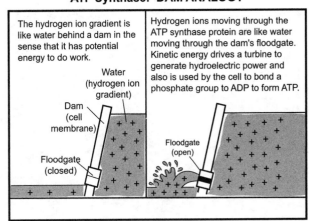

The hydrogen ion gradient is like water behind a dam in the sense that it has potential energy to do work.

Hydrogen ions moving through the ATP synthase protein are like water moving through the dam's floodgate. Kinetic energy drives a turbine to generate hydroelectric power and also is used by the cell to bond a phosphate group to ADP to form ATP.

Water (hydrogen ion gradient)

Dam (cell membrane)

Floodgate (closed)

Floodgate (open)

231

Description

Fermentation is the production of two **lactate** molecules from the breakdown of a single **glucose** molecule when no oxygen is present in the cell. This process occurs in the cytosol and yields 2 ATP. Recall that in glycolysis (see p. 226), glucose is broken down into two pyruvate molecules. When oxygen is present—**aerobic respiration**—the pyruvic acid is converted into acetyl coenzyme A, enters the citric acid cycle, and then the electron transport chain. In this case, the number of ATP produced is about 36. But when no oxygen is present, such as in skeletal muscle cells during vigorous exercise, the pyruvate is converted into lactate. This reduces ATP production to a total of only two.

The conversion of pyruvate into lactate is a type of chemical reaction called a reduction reaction. Oxidation–reduction reactions are common in metabolic pathways and are always paired. These reactions involve the transfer of electrons from one substance to another. To distinguish between these two reactions, recall OIL RIG = Oxidation Is Loss, Reduction is Gain. The NADH is oxidized to become NAD^+ and pyruvate is reduced to become lactate. NAD^+ is a carrier molecule with electrons as its cargo. It acts like a shuttle bus in that it is picking up and dropping off electrons constantly. The production of NAD^+ in the formation of lactate allows glycolysis to continue because it is needed for that process.

The buildup of lactate inside skeletal muscle cells causes soreness, which is the genesis of the phrase used by athletic trainers: "no pain, no gain." This accumulation is a problem because it tends to lower the local pH, which interferes with normal chemical reactions that have to occur. Consequently, cells have to get rid of it. As its levels increase, lactate diffuses out of cells and into the blood, from which it is taken up by liver cells. Unlike most body cells, liver cells contain special enzymes that can convert lactate back into pyruvate in the presence of oxygen. After this occurs, liver cells can convert pyruvate back into glucose. The glucose then can diffuse back into the blood, where it can be used as an energy source for other muscle cells. This chemical cycling of lactate back into glucose between muscle cells and liver cells is called the **Cori cycle**.

Roadmap

FERMENTATION

KEY

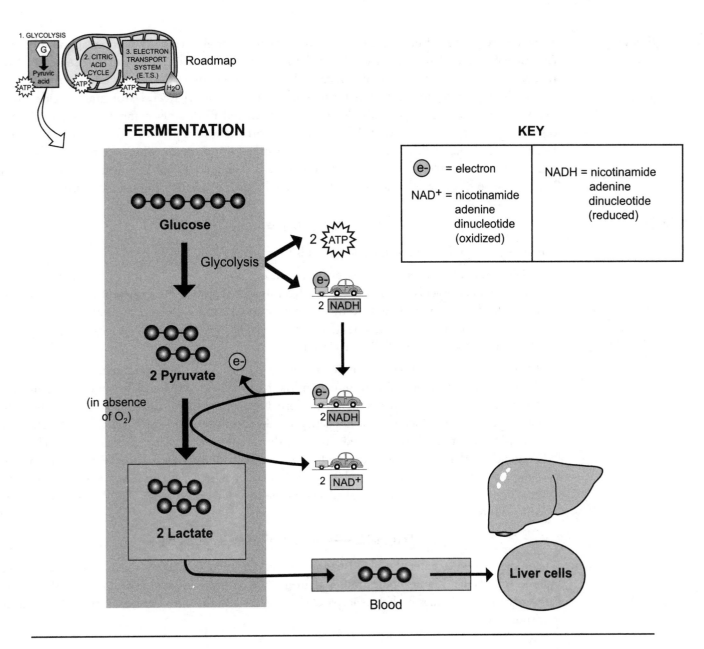

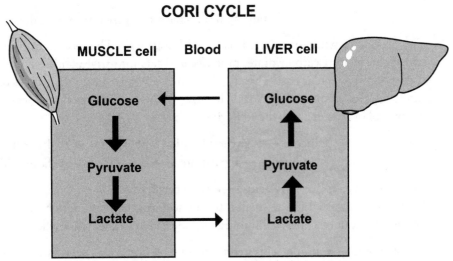

CORI CYCLE

Description

Many different types of lipids exist in the body, including **triglycerides**, **glycerophospholipids**, and **cholesterol**. Some of the stored lipids can be removed and either oxidized to be used as fuel or redeposited in other fat cells. While many cells in your body prefer to burn a carbohydrate-based fuel such as glucose, they also can burn other fuels such as triglycerides or amino acids when necessary. Alternatively, new lipids also may be synthesized from other compounds. Lipids do not dissolve in water, which makes it difficult to transport them in the blood. To solve this problem, lipids are combined with proteins to form spheres called **lipoproteins**, which are more easily transported through the blood and delivered to body cells. Triglycerides are the most common form of lipid in the body, so we will refer to this type when discussing how lipids are metabolized.

Lipolysis

(lipo = lipid; *–lysis* = breaking down)

Lipolysis is the process of breaking down lipids to produce ATP.

- **Triglycerides** are broken down via hydrolysis in the small intestine into their component parts: **glycerol** and **fatty acids**. Once inside cells, glycerol can be converted into a glycolysis intermediate—**glyceraldehyde-3-phosphate**—and then into **pyruvate**. Pyruvate is oxidized through the citric acid cycle, producing reduced coenzymes (NADH and $FADH_2$). As these coenzymes are oxidized by the electron transport system, ATP is produced.

- **Fatty acids** follow a different pathway through a process called **beta oxidation**. In this series of chemical reactions that occur in mitochondria, fatty acids are broken down into 2-carbon fragments. Each of these fragments then is converted into **acetyl CoA**, which is oxidized through the citric acid cycle, producing reduced coenzymes. As these coenzymes are oxidized by the electron transport system, ATP is produced. In total, this accounts for a large production of ATP. For example, some long-chained fatty acids produce about four times more ATP compared to the complete catabolism of a single glucose molecule. In liver cells, fatty acids are catabolized to produce a group of substances called **ketone bodies**. This process is called **ketogenesis** and it follows this pathway:

Fatty acids ⟶ acetyl CoA ⟶ **ketone bodies**

These ketone bodies enter the bloodstream and are delivered to cells. Some cells, such as cardiac muscle cells prefer ketone bodies rather than glucose to produce ATP. They accomplish this by converting ketone bodies back into acetyl CoA, which enters the citric acid cycle, then the electron transport system.

Lipogenesis

(lipo = lipid; *–genesis* = generation, birth)

The process of synthesizing lipids is called **lipogenesis**. This occurs in liver cells and adipose cells, which can convert carbohydrate, proteins, and fats into triglycerides. For example, drinking lots of soda pop on a regular basis gives the body more simple sugars than it needs, so they are converted into triglycerides and stored in fat cells. Excess simple sugars such as glucose leads to the production of glycerol or fatty acids by the following pathways:

Glucose ⟶ glyceraldehyde-3-phosphate ⟶ **glycerol** ...OR...
Glucose ⟶ pyruvate ⟶ acetyl CoA ⟶ **fatty acids**.

The excess **glycerol + fatty acids** ⟶ **triglycerides**.

Proteins are digested in the small intestine to form their component amino acids. Certain types of amino acids can be converted into triglycerides. Here is the pathway for proteins:

Protein ⟶ amino acids (*certain types*) ⟶ acetyl coA ⟶ fatty acids ⟶ **triglycerides**

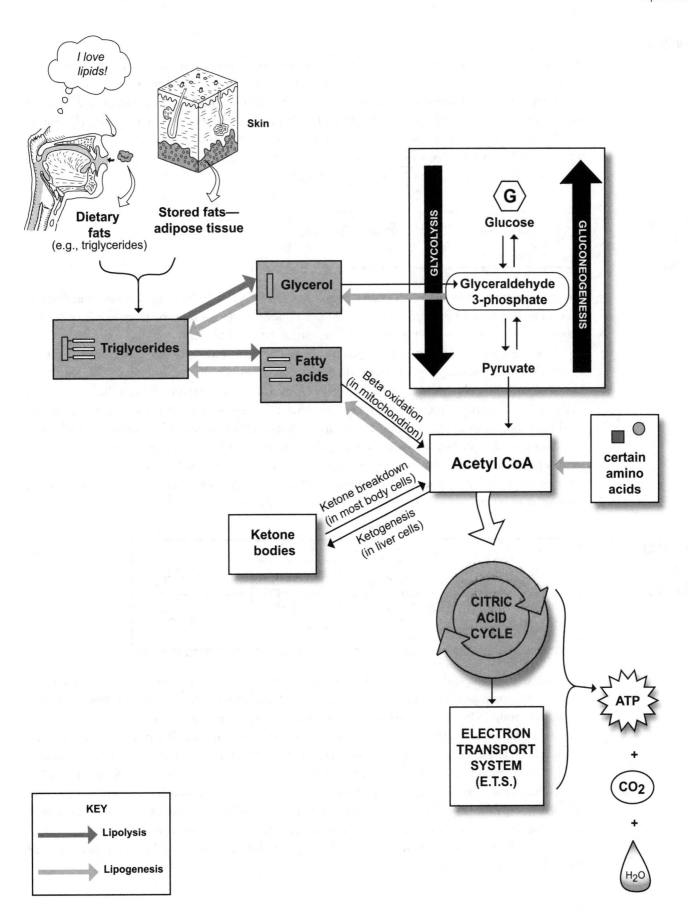

Description

Your body is made primarily of water, fat, and **protein**. A healthy adult body may contain as much as 18% protein. If you eat a chicken sandwich, the protein in the chicken will be broken down gradually into **amino acids** in the digestive tract. The amino acids then are transported into the bloodstream and delivered to body cells. All body cells need amino acids for two purposes: (1) **protein synthesis**—to make new proteins within the cells, or (2) to use as a fuel source to provide energy (to make ATP). If you ingest more protein than your body needs daily, the excess amino acids are converted into either glucose ⟨G⟩ or triglycerides ▤.

Amino Acids for Protein Synthesis

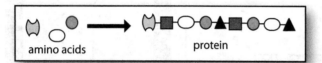

amino acids   protein

The process of protein synthesis occurs at a ribosome (see p. 197). During human growth and development, proteins have to be produced regularly and rapidly. In the adult, protein synthesis serves the purpose of replacing worn-out proteins and repairing damaged tissues.

Just as the different letters in the alphabet are used to form various words, different amino acids are used to make various proteins. There are 20 different types of amino acids (see p. 158). Your body can synthesize roughly half of them, so the others must come from our diet. Some cells, such as liver cells, are more active at protein synthesis than others. Sometimes a cell has to convert one amino acid into another to aid the process of protein synthesis. This is achieved by a process called **transamination**, in which where an α-**amino group** (NH_3^+) from an amino acid is transferred to an α-**keto acid** with the help of the enzyme transaminase. The result is that the α-keto acid is converted into a new amino acid that can be used in protein synthesis. Examples of other tissues that are active in protein synthesis are cells in the brain, skeletal muscle, and heart.

Amino Acids Used as Fuel Source

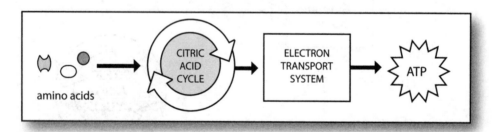

Alternatively, amino acids can be used by body cells to produce ATP by entering the citric acid cycle. For this to occur, amino acids must go through a **deamination** process whereby they lose their α-**amino group** (NH_3^+) and a hydrogen atom with the help of the enzyme deaminase. The result is that an **ammonium ion** (NH_4^+) is produced along with an α-**keto acid**. The keto acid can enter the citric acid cycle and electron transport system to produce ATP for the cell. Because ammonium ions (NH_4^+) in high quantities are toxic to cells, they are further converted into a waste product called urea, a harmless substance. This occurs in liver cells when ammonium ions and carbon dioxide enter a biochemical pathway called the urea cycle, in which urea is the primary side product. The liver is the major site for these deaminations because the liver cells have the proper enzymes to do the job. Urea normally travels through the blood and into the kidneys, where it is filtered out and excreted from the body as part of the urine.

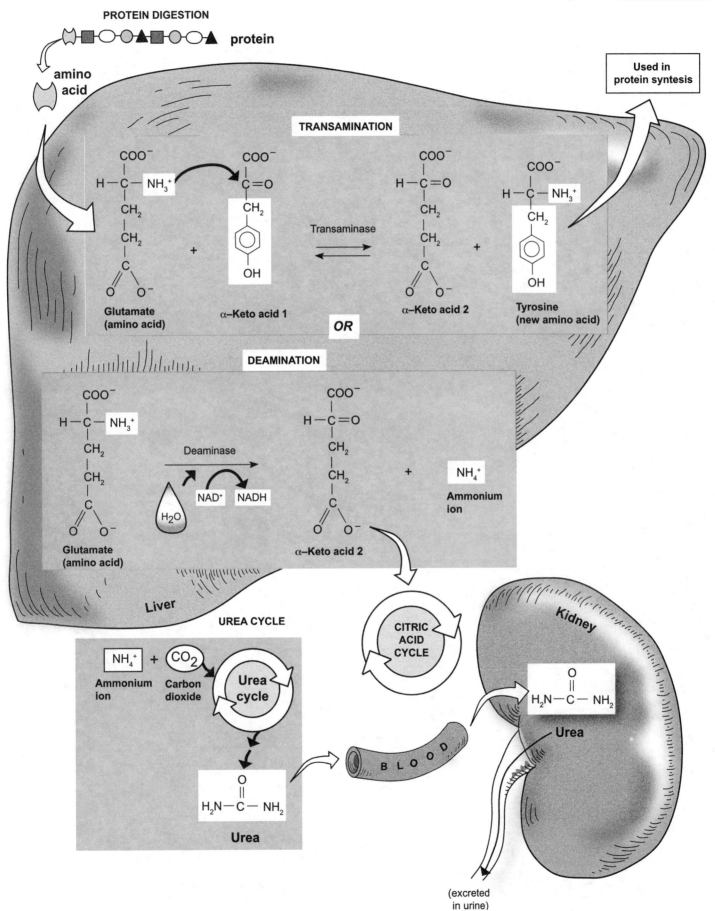

PROTEIN DIGESTION

protein

amino acid

TRANSAMINATION

Transaminase

Glutamate (amino acid)

α–Keto acid 1

OR

α–Keto acid 2

Tyrosine (new amino acid)

Used in protein syntesis

DEAMINATION

Deaminase

NAD⁺   NADH

H_2O

Glutamate (amino acid)

α–Keto acid 2

+

NH_4^+
Ammonium ion

Liver

UREA CYCLE

NH_4^+ + CO_2

Ammonium ion   Carbon dioxide

Urea cycle

$H_2N-C-NH_2$

Urea

CITRIC ACID CYCLE

Kidney

BLOOD

$H_2N-C-NH_2$

Urea

(excreted in urine)

Glossary

A

acetyl CoA ass-EE-til CoA two-carbon acetyl unit bonded to coenzyme A

acid ASS-id substance that forms hydrogen ions when dissolved in water

acidosis ass-i-DOH-sis physiological condition in which the blood pH value is less than 7.35

actinides ak-tuh-NYDEZ group of 15 metallic chemical elements with atomic numbers 89–103, begins with actinium

activation energy ak-tih-VAY-shun EN-er-gee energy needed to initiate a chemical reaction

active site AK-tiv syte site within an enzyme that binds a substrate or substrates and catalyzes the reaction

adenosine diphosphate (ADP) ah-DEN-oh-seen dye-FAHS-fayt molecule composed of adenine, ribose, and two phosphate groups

adenosine triphosphatase (ATPase) ah-DEN-oh-seen try-FOS-fuh-tayz enzyme that catalyzes the conversion of ATP into ADP

adenosine triphosphate (ATP) ah-DEN-oh-seen try-FAHS-fayt energy currency of a cell; composed of an adenine base, a ribose sugar, and three phosphate groups

aerobic air-OH-bik any reaction or process that requires oxygen

alcohols AL-koh-hawls class of organic compounds than contains a hydroxyl (-OH) group bonded to a carbon atom

aldehydes AL-dah-hydes class of organic compounds that contains a terminal carbonyl group(s) bonded to a hydrogen atom(s)

alkaline earth metals AL-kah-lye metallic elements found in Group 2A of the periodic table; two electrons in outer shell; very reactive

alkali metals AL-kah-lye metallic elements found in Group 1A of the periodic table; highly reactive and softer than most metals

alkaloids AL-kah-loyds biologically active amines synthesized by plants to ward off insects and animals

alkalosis al-kah-LOH-sis physiological condition in which the blood pH value is greater than 7.45

alkanes AL-kanes class of organic compounds that contain only carbon and hydrogen atoms and only form single bonds

alkenes AL-keens class of organic compounds that contain carbon-carbon double bonds

alkyl group AL-kul an alkane minus one hydrogen atom

alkynes AL-kines class of organic compounds that contain carbon-carbon triple bonds

alpha helix AL-fah HEE-lix secondary level of protein structure in which a hydrogen bond connects the NH of one peptide bond with the C=O of a peptide bond later in the chain to form a coiled or spiral conformation

alpha particle AL-fuh PAHR-ti-kuhl radioactive particle that contains 2 protons and 2 neutrons

amides AM-eyeds class of organic compounds derived from ammonia or an amine by replacement of an atom of hydrogen with an acyl group

amines AM-eens class of organic compounds derived from ammonia by replacement of one or more hydrogen atoms with organic groups

amino acids ah-MEE-no ASS-ids class of biochemicals that contains both an amino group and a carboxyl group bonded to the same carbon; building blocks for making proteins

amino group ah-MEE-no functional group composed of a nitrogen bonded to two hydrogens

amylase AM-eh-layz enzyme that breaks down polysaccharides; produced by salivary glands and the pancreas

amylopection am-uh-loh-PEK-tin branched form of plant starch composed of glucose molecules

amylose AM-uh-lohs unbranched form of plant starch composed of α-D-glucose molecules connected by a α-1,4 glycosidic bonds in a continuous chain

anabolic reactions an-ah-BOL-ik ree-AK-shuns chemical reactions in which simpler substances combine to form more complex substances, usually requiring energy

anaerobic an-air-OH-bik any reaction or process that does not require oxygen

anion AN-eye-on negatively charged ion

anticodon an-tee-KOH-don triplet base sequence found on transfer RNA (tRNA) that is complementary to the mRNA codon

aqueous solution AY-kwee-uhs suh-LOO-shun any solution with water as its solvent

aromatic compounds air-ah-MAT-ik KOM-pounds compounds whose molecular structure includes one or more planar rings of atoms in which electrons are shared equally

atmospheric pressure (A.P.) at-muss-FER-ik pressure exerted by the atmosphere

atom AT-om smallest unit of a chemical element that displays the properties of that element

atomic energy ah-TOM-ik form of energy produced by nuclear fission

atomic mass ah-TOM-ik mass of a single atom

atomic number ah-TOM-ik equal to the number of protons in an atom

atomic orbital OR-bit-al region in space around an atom's nucleus in which an electron is most likely to be found

atomic symbol ah-TOM-ik abbreviation used to indicate an element

ATP *see* adenosine triphosphate

ATPase AY-TEE-PEE-ayz *see* adenosine triphosphatase

ATP synthase SIN-thays enzyme in the mitochondrion that synthesizes ATP from ADP and inorganic phosphate

Avogadro's number ah-vuh-GAH-droh's NUHM-ber number of particles in one mole, equal to 6.02×10^{23}

B

bases Baysez substances that form hydroxide ions when dissolved in water

becquerel (Bq) bek-uh-REL SI unit for measuring emitted radiation

benzene BEN-zeen an aromatic hydrocarbon characterized by a ring of six carbons with each bonded to a hydrogen

beta oxidation BAY-tah ahk-si-DAY-shun degradation of fatty acids that removes two-carbon segments from the end of the fatty acid

beta particle BAY-tuh PAHR-ti-kuhl radioactive particle that is an electron

beta-pleated sheet BAY-tah secondary level of protein structure created by hydrogen bonds between lateral sections of the polypeptide

boiling BOY-el-ling phase transition from liquid to gas

boiling point BOY-el-ling the temperature at which the vapor pressure of a liquid equals the atmospheric pressure surrounding the liquid

Boyle's law BOILS law gas law that states the volume of a gas is inversely proportional to its pressure when there is no change in the temperature or the amount of the gas

buffers BUFF-erz solution that prevents change in pH when an acid or base is added to it

C

calorie KAL-or-ee amount of heat needed to raise the temperature of 1 g of water by 1°C

carbohydrates kar-boh-HYE-drayts polymers made from sugars; class of biochemicals referred to as simple or complex sugars

carbonyl group KAR-buh-nil functional group of carbohydrates composed of a carbon double bonded to an oxygen

carboxyl group kar-BOK-sil functional group of carbohydrates composed of a carbonyl group bonded to hydroxyl group

carboxylic acid kar-BOK-sil-ik ASS-id class of organic compounds that contain the carboxyl functional group

catabolic reactions kat-ah-BOL-ik ree-AK-shuns metabolic reactions that break down complex substances into simpler ones with the release of energy

catalysts KAT-ah-lists substances that accelerate the rate of a chemical reaction by decreasing the activation energy without themselves being consumed

cation KAT-eye-on positively charged ion

cellular respiration sel-yuh-ler res-puh-RAY-shun biochemical process involving the oxidation of glucose in a cell that has three parts: (1) glycolysis, (2) citric acid cycle, (3) electron transport system

cellulose SEL-yoo-lohs carbohydrate that is a polymer composed of glucose units; main component of the cell walls of most plants; insoluble in water

Celsius (C) scale SEL-see-us temperature scale on which the freezing point of water is 0°C and the boiling point of water is 100°C

centimeter (cm) SEN-ti-mee-ter unit of length equal to 100th of a meter

Charles' law charlz law gas law that states the volume of a gas is directly related to the temperature when there is no change in pressure or the amount of gas

chemicals KEM-ih-kalz any substances used in or produced by a chemical process involving changes to atoms or molecules

chemical change KEM-ih-kuhl change change that involves a chemical reaction in which one or more new substances are produced

chemical equation KEM-ih-kal symbolic representation of a chemical reaction

chemical formula FOR-myoo-lah expression of the constituents of a compound by symbols and figures

chemical reaction KEM-ih-kal process by which a chemical change occurs

chemical symbols KEM-ih-kal abbreviation that represents the name of an element

chemistry KEM-ih-stree study of substances and the changes they undergo

cholesterol koh-LESS-ter-ol steroid compound produced by animals, found in their cell membranes

citric acid cycle SIT-rik ASS-id SYE-kul cyclic, aerobic chemical pathway consisting of eight reactions that begins with the formation of citric acid and results in the formation of oxaloacetic acid

codon KOH-don triplet base sequence found on messenger RNA (mRNA)

coenzymes koh-EN-zymes nonprotein biochemicals that take part in enzymatically catalyzed reactions

cofactors KOH-fak-ters various organic or inorganic substances necessary to the function of an enzyme

collagen KAHL-ah-jen structural protein consisting of three polypeptide chains arranged in a triple helix that are bundled together in fibers

colloids KOL-oyds heterogeneous mixtures that do not separate or settle out; particle size is small enough to pass through filters, but too large to pass through semipermeable membranes

combination reaction kom-buh-NAY-shun ree-AK-shun type of chemical reaction where two or more reactants combine to form a single product

compound KOM-pound substance composed of two or more atoms of elements in a fixed proportion

concentration (*of a solution*) kahn-sen-TRAY-shun amount of solute dissolved in a given quantity of solution

condensation kon-den-SAY-shun changing from a gas state to a liquid state through removal of heat

covalent bond koh-VAYL-ent bond chemical bond formed by two atoms sharing one or more pairs of electrons

crenation kreh-NAY-shun process resulting from osmosis in which cells, in a hypertonic solution, undergo shrinkage

curie KYOO-ree convention unit for measuring emitted radiation

cytoplasm SYE-toh-plaz-em gel-like material between the nucleus and the cell membrane; includes cytosol and organelles

cytosol SYE-toh-sawl water soluble portion of the cytoplasm

D

Dalton's law DAL-tenz law gas law that states the total pressure exerted by a mixture of gases is the sum of the pressure exerted by each individual gas

decarboxylation dee-kar-bok-sah-LAY-shun removal of a carboxyl group from a chemical compound

decomposition reactions dee-kom-poh-SIH-shun ree-AK-shuns type of chemical reaction where a compound breaks down into its component parts

dehydration dee-hye-DRAY-shun process of removing water from a substance or compound

dehydration synthesis dee-hye-DRAY-shun SIN-thuh-sis process of removing a water molecule to form a new covalent bond between two monomers

denaturation dee-nayt-chur-AY-shun to treat (a protein or the like) by chemical or physical means so as to alter its original three-dimensional state resulting in loss of function

density DEN-sih-tee the mass of a substance divided by its volume

deoxyribonucleic acid (DNA) dee-ok-see-rye-boh-noo-KLAY-ik ASS-id nucleic acid that is the genetic material determining the makeup of all living cells and many viruses

dephosphorylation dee-FOS-for-ih-lay-shun process of removing an inorganic phosphate group from a compound

deposition dep-oh-ZISH-un changing from a gas state directly into a solid state through the removal of heat

dialysis dye-AL-i-sis separation of large solutes from small solutes in a solution by allowing the latter to pass through a semipermeable membrane

diffusion dih-FYOO-shun net movement of substances from an area of high concentration to an area of low concentration

digestion dye-JES-chun process that breaks down large, complex molecules into simpler ones

dipeptide dih-PEP-tyde two amino acids covalently bonded together

disaccharides dye-SAK-ah-rydes carbohydrate consisting of two monosaccharides bonded together; a double sugar

disulfide bond dye-SUL-fyde covalent bond between the sulfur atoms on two cysteine molecules in a peptide or a protein

dipole DYE-pohl separation of charge between two covalently bonded atoms

DNA see deoxyribonucleic acid

double covalent bond DUHB-uhl koh-VAYL-ent bond chemical bond between two atoms with two shared pairs of electrons

double helix HEE-lix pair of parallel helices intertwined around a common axis

double replacement reaction reaction that involves the exchange of ions between two compounds

E

electrolyte eh-LEK-troh-lyte chemical compound that forms ions in solution

electron ee-LEK-tron negatively charged subatomic particle located outside the nucleus of an atom

electron cloud ee-LEK-tron kloud region around the nucleus of an atom where electrons are located

electron-dot structures ee-LEK-tron dot struhk-chers visual shorthand method to depict valence electrons

electron transport system (E.T.S.) ee-LEK-tron TRANS-port SIS-tem components of the final sequence of reactions in biochemical oxidation

electronegativity eh-lek-troh-NEG-uh-tiv-ih-tee relative ability of an atom to attract bonding electrons to itself in a chemical bond

elements EL-ah-muhnt substance that cannot be broken down into simpler substances by chemical means

endothermic reactions en-doh-THUR-mik ree-AK-shuns reactions that absorb heat from their surroundings

energy EN-er-jee ability to do work or transfer heat

enzymes EN-zymes organic molecules—usually proteins—that catalyze biological reactions

enzyme inhibitors EN-zyme in-HIB-i-torz substances that cause enzymes to lose catalytic activity

enzyme-substrate (ES) complex EN-zyme-SUB-strayt KOM-pleks enzyme bound to a substrate in an enzyme-catalyzed reaction

esterification eh-ster-uh-fih-CAY-shun reaction in which a carboxylic acid reacts with an alcohol to produce an ester and water

esters ES-terz class of organic compounds that contain a carboxyl group bonded to an oxygen that is bonded to carbon

ethers EE-therz class of organic compounds that contains an oxygen atom bonded to two carbon atoms

exothermic reactions ek-soh-THUR-mik ree-AK-shuns reactions that release heat to their surroundings

experiment ik-SPER-uh-muhnt operation or procedure designed to test the validity of a hypothesis

F

fat-soluble vitamins FAT SOL-yoo-bul vye-tuh-minz vitamins that are not soluble in water and can be stored in body fat

fatty acids ASS-ids long-chain carboxylic acids used as the building blocks of triacylglycerols and glycerophospholipids

fermentation fur-men-TAY-shun process of producing two lactate molecules from the breakdown of one glucose molecule when no oxygen is present in the cell

filtration fil-TRAY-shun movement of water and solutes through a membrane by a higher hydrostatic pressure on one side of the membrane

fission FISH-uhn splitting of atomic nuclei resulting in the release of large amounts of energy

flavin adenine dinucleotide (FAD) FLAY-vin AD-eh-neen dye-NOO-klee-oh-tyde coenzyme derived from riboflavin that serves as an electron and proton carrier

formula FOR-myoo-lah see chemical formula

freezing FREE-zing changing from a liquid state to a solid state through the removal of heat

fructose FRUK-tohs monosaccharide found in many fruits and honey

functional groups FUNK-shen-all small clusters of bonded atoms responsible for the physical and chemical characteristics of a class of organic compounds

fusion FYOO-shen reaction in which large amounts of energy are released when small nuclei combine to form larger nuclei

G

galactose gah-LAK-tohs monosaccharide typically found as a component of the disaccharide lactose

gamma rays GAM-ah rayz rays that deliver high-energy radiation

gas state of matter characterized by no definite shape or volume

Gay-Lussac's law GAY lah-SAK gas law that states the pressure is directly proportional to the temperature in Kelvin when there is no change in the volume and amount of the gas

genetic code jeh-NET-ik biochemical basis of heredity consisting of nucleotide codons translated into specific amino acids

globular proteins GLOB-yoo-lur PROH-teens proteins that acquire a compact shape from attractions between the R group of the amino acid residues in the protein

glucose GLOO-kohs monosaccharide sugar occurring widely in most plant and animal tissues

glycerol GLIS-uh-rawl alcohol that fatty acid chains bond with to form a triacylglycerol

glycogen GLYE-koh-jen branched glucose α-D-polysaccharide that is the main form of carbohydrate storage in animals

glycolysis glye-KOHL-i-sis first in the series of chemical reactions of glucose catabolism

gram (g) metric unit used for mass

Gray (Gy) SI unit for measuring an absorbed dose of radiation

group vertical column in the periodic table that contains elements having similar physical and chemical properties

group numbers number that appears at the top of each vertical column (group) in the periodic table; indicates the number of electrons in the outermost energy level

H

half–life length of time it takes for half of a radioactive sample to decay

halogens HAL-uh-jens nonmetal elements found in Group 7A of the periodic table

heat heet transfer of energy between two bodies of different temperature

hemodialysis hee-moh-dye-AL-i-sis use of dialysis to separate waste products from blood

hemolysis hee-MOL-i-sis destruction of red blood cells in a hypotonic solution due to an increase in fluid volume

Henry's law gas law that states that the solubility of a gas, at a given temperature, is proportional to the pressure of the gas above a liquid

heterogeneous mixture het-er-oh-JEE-nee-us any combination of substances that does not have uniform composition and properties

homogeneous mixture hoh-moh-JEE-nee-us any combination of substances that has uniform composition and properties

hydration hye-DRAY-shun process of surrounding dissolved ions or molecules by water molecules

hydrocarbons hye-droh-KAR-bons any class of organic compounds containing only hydrogen and carbon

hydrogen bonds HYE-droh-jen electrostatic interaction between a hydrogen atom covalently bonded to oxygen, nitrogen, or fluorine and another oxygen, nitrogen, or fluorine atom

hydrolysis hye-DROHL-i-sis process of breaking a chemical bond in a molecule with the addition of water

hydrophilic interactions (*in proteins*) hye-droh-FIL-ik attractions between the external aqueous environment and amino acids that have polar or ionized side chains

hydrophobic interactions (*in proteins*) hye-droh-FOH-bik interactions between two nonpolar R groups

hydroxyl group hye-DROK-sul functional group of carbohydrates composed of an oxygen bonded to a hydrogen

hypertonic hye-per-TON-ik a solution having a higher osmotic pressure than another solution

hypothesis hye-POTH-eh-sis statement that proposes a tentative explanation for a set of observations

hypotonic hye-poh-TON-ik a solution having a lower osmotic pressure than another solution

I

inhibitors in-HIB-i-torz *see* enzyme inhibitors

inorganic compounds in-or-GAN-ik KOM-pounds compounds that do not contain hydrocarbon groups

International System of Units (SI) system of units known as the modern metric system

International Union of Pure and Applied Chemistry (IUPAC) international group of scientists that developed a set of rules for naming organic and inorganic chemical compounds

ion EYE-on atom or group of atoms having an electrical charge as a result of losing or gaining electrons

ionic bond eye-ON-ik electrostatic force that holds ions together in an ionic compound

isomers EYE-soh-merz different substances that have the same chemical formula

isotonic eye-soh-TON-ik a solution having the same osmotic pressure as another solution

isotopes EYE-so-tohp atoms with the same atomic number but different atomic masses

IUPAC *see* International Union of Pure and Applied Chemistry

J

joule (J) jool unit of energy; SI unit of heat energy

K

Kelvin (K) scale KEL-vin temperature scale on which absolute zero (-273.15°C) is 0 K, the freezing point of water is 273.15 K and the boiling point of water is 373.15 K

ketone bodies KEE-tohn product of ketogenesis: acetoacetate, beta-hydroxybutyric acid, or acetone

ketones KEE-tonz monosaccharides class of organic compounds containing a carbonyl group bonded to two alkyl groups

ketosis kee-TOH-sis condition in which high levels of ketone bodies cannot be metabolized resulting in a reduced blood pH

kilocalorie (kcal) KIL-oh-kal-oh-ree amount of heat equal to 1,000 calories

kilogram (kg) KIL-oh-gram unit of mass equal to 1,000 grams; standard SI unit of mass

kinetic energy kih-NET-ik EN-er-jee energy associated with motion

L

lactose LAK-tohs disaccharide composed of glucose and galactose

lanthanides lan-thuh-NYDZ group of 15 metallic chemical elements with atomic numbers 57–71 known as the rare earth elements, begins with lanthanum

lipid–soluble vitamins *see* fat-soluble vitamins

lipids LIP-ids class of water-insoluble biochemicals

lipogenesis lye-poh-JEN-ih-sis process of synthesizing lipids

lipolysis li-POL-ih-sis process of breaking down lipids to produce ATP

liquid LIK-wid state of matter in which a substance exhibits a characteristic readiness to flow, little or no tendency to disperse, and is relatively incompressible

liter (L) LEE-ter metric unit used in measurements of volume

M

maltose MAL-tohs disaccharide composed of two glucose molecules

mass measure of the amount of matter in an object

mass number sum of the number of protons and neutrons in an atom

matter anything that has mass and occupies space

melting MEL-ting changing from a solid state to a liquid state through the addition of heat

melting point MEL-ting temperature at which the particles in the solid gain sufficient energy to overcome the attractive forces that hold them together

messenger RNA (mRNA) sequence of RNA nucleotides that specifies the order of amino acids in a protein

metabolism meh-TAB-oh-liz-em sum total of chemical reactions that take place in an organism

metal MEH-tal element with high electrical conductivity, luster, and malleability

metalloid MET-al-loyd element that exhibits some properties typical of metals and other properties characteristic of nonmetals

meter (m) MEE-ter metric unit of length; 100 centimeters

metric system MET-rik SIS-tem decimal system of measurement based on the meter as a unit of length, the kilogram as a unit of mass, and the second as a unit of time

milliliter (mL) MIL-ah-lee-ter unit of volume equal to one thousandth of a liter

mitochondria my-toh-KON-dree-ah cellular organelles that are the site of aerobic cellular respiration

mixture MIKS-chur two or more substances that are physically mixed, but not chemically combined

molar mass MOH-ler mas mass of one mole of any substance expressed in units of grams/mole

molarity mohl-LAR-ih-tee number of moles of solute per liter of solution

mole (mol) mohl unit of measurement equal to 6.02×10^{23} particles of a substance

molecule MOH-eh-kyool compound consisting of two or more atoms held together by chemical bonds

monomer MON-oh-meer basic unit of a polymer

monosaccharide mon-oh-SAK-ah-ryde carbohydrate that contains a single carbonyl group and two or more hydroxyl groups

mutation myoo-TAY-shun change in DNA that can cause subsequent changes in an organism that can be transmitted genetically

N

neutralization noo-trahl-ih-ZAY-shun reaction of an acid and a base

neutron NOO-tron neutrally charged subatomic particle in the nucleus of an atom

nicotinamide adenine dinucleotide (NAD⁺) nik-oh-TIN-ah-myde AD-eh-neen dye-NOO-klee-oh-tyde coenzyme derived from nicotinamide that serves as an electron and proton carrier

noble gas NO-bul elements in Group 8A of the periodic table

nonelectrolyte non-ee-LEK-troh-lyte compound that does not form ions in solution and does not conduct an electrical current

nonmetal non-MEH-tal elements with little or no luster and are poor conductors of heat or electricity

nonpolar molecule non-POH-lar MOH-eh-kyool molecule that has only nonpolar bonds or in which the bond dipoles cancel

nuclear fission NOO-klee-er FISH-uhn process of splitting a radioactive nucleus to produce new, smaller nuclei

nucleic acids noo-KLAY-ik ASS-ids class of biochemicals that are polymers of nucleotides

nucleoside NOO-klee-oh-syde purine or pyrimidine base bonded to a sugar, typically ribose or deoxyribose

nucleotides NOO-klee-oh-tides monomers made of a sugar, phosphate, and a nitrogen base

nucleus NOO-klee-UHS center of an atom that contains protons and neutrons

O

octet ok-TET group of eight valence electrons surrounding an atom

octet rule ok-TET general principle in chemistry whereby atoms other than hydrogen usually form bonds until each atom has eight valence electrons

oil oyl triacylglycerol that is a liquid at room temperature

orbital OR-bit-al *see* atomic orbital

organic compounds Or-GAN-ik KOM-pounds compounds made of carbon that typically have covalent bonds, nonpolar molecules, low melting point and boiling point, are insoluble in water, and flammable

osmosis os-MOH-sis movement of solvent molecules across a semipermeable membrane toward the solution containing the higher solute concentration

osmotic pressure os-MOT-ik PRESH-ur water pressure that results from a concentration gradient

oxidation AHK-sih-day-shun loss of one or more electrons from an atom or molecule

oxidation–reduction reaction (redox reactions) type of chemical reaction that involves the transfer of electrons between a pair of substances

oxidative phosphorylation ahk-si-DAY-tiv fos-for-i-LAY-shun in cell metabolism, the process by which respiratory enzymes in mitochondria synthesize ATP from ADP and inorganic phosphate

oxidizing agent OK-si-dyz-ing AY-juhnt electron receiver in an oxidation–reduction reaction

P

partial pressure PAR-shal PRESH-ur pressure exerted by a single gas in a mixture

peptide PEP-tyde compound made of two or more amino acids bonded by peptide bonds

peptide bond PEP-tyde amide bond formed between two amino acids

periodic table peer-ee-OD-ik chart illustrating the elements in periods and rows, arranged by increasing atomic number

pH the negative logarithm of the molar concentration of hydronium ion (H_3O^+)

pH scale scale for determining if a solution is acidic or basic, ranging from 0 – 14 with 7 as the neutral point; values < 7 are acidic, values > 7 are basic

phenol FEE-nohl organic compound where a hydrogen in a benzene ring is replaced by a hydroxyl group

phospholipid bilayer fos-fo-LIP-id BYE-lay-er two-layered arrangement of phospholipids that form a cell membrane

phosphorylation FOS-fer-ih-lay-shun process of adding an inorganic phosphate group to another compound

physical change modification in a substance that does not alter its composition

physical properties characteristics that can be observed or measured without changing the composition of a substance

pKa symbol for the acid dissociation constant

polar molecule POH-lar MOHL-eh-kyool molecule containing polar bonds where the sum of all the bonds' dipole moments is not zero

polymer PAHL-i-mer large molecule made up of smaller molecules, monomers, bonded together in sequence

polypeptide pol-ee-PEP-tyde chain of many amino acids covalently bonded together

polysaccharide pahl-ee-SAK-ah-ryde carbo-hydrate polymer composed of more than 10 monosaccharide units

positron POZ-i-tron radioactive particle that is a positively charged Beta particle

potential energy poh-TEN-shul EN-er-jee energy that matter possesses

pressure PRESH-ur force applied to a given area

primary structure (*in proteins*) PRYE-mar-ee STRUK-chur sequence of amino acids in a protein

products PROD-ukts final substances formed in a chemical reaction, written to the right of the arrow

protein PROH-teen class of polymeric bio-chemicals composed of long chains of amino acids

protein synthesis PROH-teen SIN-thuh-sis the process of creating new proteins within a cell

proton PROH-ton positively charged sub-atomic particle located in the nucleus of an atom

proton pumps PROH-ton membrane proteins capable of moving protons across the mem-brane of a cell or organelle

pyruvate pye-ROO-vayt final product produced at the end of glycolysis

Q

quaternary structure (*in proteins*) KWAH-ter-nair-ee STRUK-chur structural level of a biologi-cally active protein that consists of two or more polypeptide chains

R

rad rad conventional unit for measuring an absorbed dose of radiation

radiation ray-dee-AY-shun energy emission from an unstable nucleus in the form of particles or rays

radioactive decay ray-dee-oh-AK-tiv process by which unstable nuclei break down with the release of radiation

radioisotope ray-dee-oh-EYE-soh-tope isotope that emits radiation

reactants ree-AK-tuhnts starting substances in a chemical reaction written to the left of the arrow

redox reactions REE-doks ree-AK-shuns *see* oxidation–reduction reactions

reducing agent ri-DOOS-ing AY-juhnt electron donor in an oxidation–reduction reaction

reduction rih-DUK-shun gain of one or more electrons from an atom or molecule

rem conventional unit for measuring biological risk from exposure to radiation

replacement reaction ri-PLAYS-muhnt ree-AK-shun type of chemical reaction where com-ponents recombine to form new products

replication rep-lih-KAY-shun process by which double-stranded DNA makes copies of itself

retrovirus ret-troh-VYE-russ RNA virus that incorporates its genetic material into the genome of a host cell as a means to propagate

ribonucleic acid (RNA) rye-boh-noo-KLAY-ik AHS-id type of nucleic acid that has many different classes, often involved in protein synthesis

ribosomal RNA (rRNA) rye-boh-SOHM-al RNA that is a permanent structural part of a ribosome

RNA *see* ribonucleic acid

S

saturated fatty acids SATCH-yoo-ray-ted FAT-tee ASS-ids fatty acids containing no double bonds between carbon atoms

saturated solution SATCH-yoo-ray-ted suh-loo-shun solution that contains the maximum amount of dissolved solute

scientific notation shorthand method for writing out very large or very small numbers

secondary structure (*in proteins*) SEK-on-dair-ee STRUK-chur arrangement of a polypeptide into a regular alpha helix, beta structure, or random coil configuration

semipermeable membrane sem-ee-PUR-mee-ah-bul MEM-brayn membrane that allows the passage of solvent molecules and very small solute particles but not large solute molecules

SI *see* International System of Units

Sievert (Sv) SEE-vert SI unit for measuring biological risk from exposure to radiation

single covalent bond single koh-VAYL-ent bond chemical bond between two atoms with one shared pair of electrons

single replacement reactions reaction in which an atom of one element replaces the atom of another element

solid SOL-id state of matter that has a definite shape and volume

solubility sol-yoo-BIL-ih-tee the maximum amount of solute that can be dissolved in a solvent at a given temperature

solute SOL-oot dissolved substance(s) in a solution

solution suh-LOO-shun homogeneous mixture composed of a solute dissolved in a solvent

solvent SOL-vent most abundant component of a solution that dissolves the solute

steroid STAYR-oyd class of lipids characterized by its four fused carbon rings structure

subatomic particle sub-eh-TOM-ik particle that makes up an atom

sublimation sub-luh-MAY-shun changing from a solid state, through the addition of heat, without forming a liquid first

substrate sub-STRAYT substance on which an enzyme acts

sucrose SOO-krohs disaccharide composed of glucose and fructose

suspension suh-SPEN-shun mixture of two or more components with particles that can be evenly distributed by mechanical means but will eventually settle out

synthesis SIN-the-sis process of producing a compound by a chemical reaction or series of reactions

T

temperature TEM-per-ah-choor measure of the kinetic energy of a given matter

temperature and pressure law *see* Gay-Lussac's law

temperature and volume law *see* Charles' law

tertiary structure (*in proteins*) TUR-she-AIR-ee STRUK-chur helixes or beta structures of a polypeptide that are folded or arranged into a three-dimensional configuration; stabilized by the interactions of R groups

transamination trans-am-ih-NAY-shun transfer of an amino group from one compound to another

transcription trans-KRIP-shun transfer of genetic information from DNA through the formation of RNA

transfer RNA (tRNA) RNA molecules that transfer amino acids to ribosomes during protein synthesis

translation trans-LAY-shun process by which a messenger RNA molecule is converted into a corresponding linear sequence of amino acids on a ribosome to form a protein

triacylglycerol (**triglyceride**) try-as-ill-GLIS-er-ol (try-GLIS-uh-ryde) class of lipids containing three fatty acid chains bonded to a glycerol backbone

tripeptide try-PEP-tyde three amino acids covalently bonded together

triple covalent bond TRIP-ul koh-VAYL-ent bond chemical bond between two atoms with three shared pairs of electrons

U

unsaturated fatty acids un-SATCH-yoo-ray-ted FAT-tee ASS-ids fatty acids containing one or more double bonds between carbon atoms

unsaturated hydrocarbon un-SATCH-yoo-ray-ted hye-droh-KAR-bon hydrocarbon with double or triple carbon-carbon bonds

unsaturated solution Un-SATCH-yoo-ray-ted suh-LOO-shun solution that contains less solute than can be dissolved

urea cycle YOOR-ee-ah SYE-kul series of reactions during which ammonium ions from the degradation of amino acids are converted to urea

V

valence electrons VAY-lenz ee-LEK-trons electrons of an atom that are involved in chemical bonding

vaporization vay-per-uh-ZAY-shun changing from a liquid state to a gas state through the addition of heat

viruses VYE-russ-es microscopic particles that contain a nucleic acid; bound by a protein coat and sometimes a lipoprotein coat

vitamins VYE-tuh-minz nonenergy yielding biochemicals that are essential for normal health and growth

volume VOL-yoom amount of space occupied by a substance

W

water–soluble vitamins WAW-ter SOL-yoo-bul vye-tuh-minz vitamins that form a solution when mixed with water

Z

zwitterion tsvit-er-ahy-uhn dipolar molecule that has an equal amount of positive and negative charge giving it a net charge of zero

Index